Monitoring and Assessment of Structures

T0187814

Monitoring and Assessment of Structures

G. S. T. Armer

CRC Press
Taylor & Francis Group
Boca Raton London New York

CRC Press is an imprint of the
Taylor & Francis Group, an **informa** business

CRC Press
Taylor & Francis Group
6000 Broken Sound Parkway NW, Suite 300
Boca Raton, FL 33487-2742

First issued in paperback 2019

© 2001 G. S. T. Armer
CRC Press is an imprint of Taylor & Francis Group, an Informa business

No claim to original U.S. Government works

ISBN-13: 978-0-419-23770-9 (hbk)
ISBN-13: 978-0-367-86638-9 (pbk)

This book contains information obtained from authentic and highly regarded sources. Reasonable efforts have been made to publish reliable data and information, but the author and publisher cannot assume responsibility for the validity of all materials or the consequences of their use. The authors and publishers have attempted to trace the copyright holders of all material reproduced in this publication and apologize to copyright holders if permission to publish in this form has not been obtained. If any copyright material has not been acknowledged please write and let us know so we may rectify in any future reprint.

Except as permitted under U.S. Copyright Law, no part of this book may be reprinted, reproduced, transmitted, or utilized in any form by any electronic, mechanical, or other means, now known or hereafter invented, including photocopying, microfilming, and recording, or in any information storage or retrieval system, without written permission from the publishers.

Trademark Notice: Product or corporate names may be trademarks or registered trademarks, and are used only for identification and explanation without intent to infringe.

Typeset in Sabon by Lucid Digital

British Library Cataloguing in Publication Data
A catalogue record for this book is available from the British Library

Library of Congress Cataloging in Publication Data
Armer, G. S. T.
   Monitoring and assessment of structures/G.S.T. Armer.
      p. cm.
   Includes bibliographical references and index.
      (alk. paper)
   1. Structural analysis (Engineering) 2. Buildings—Testing. I. Title.

TA654 .A75 2001
624.1'71—dc21   2001020140

Visit the Taylor & Francis Web site at
http://www.taylorandfrancis.com

and the CRC Press Web site at
http://www.crcpress.com

Contents

Contributors

Graham Armer
Principal
GA Consultants, UK

David K. Cook
Associate Director
Mott MacDonald, UK

Brian R. Ellis
Technical Director
Centre for Structural Engineering
BRE, UK

Professor Dr-Ing. Klaus Steffens
Hochschule Bremen, Germany

Donald Stirling
City University, UK

Preface

The last ten years have seen an amazing improvement in the development and availability of techniques and instrumentation suitable to monitor and assess the performance and condition of structures. This phenomenon has been accompanied by a similar improvement in the accessibility low-cost computer hardware necessary for the storage and analysis of large quantities of data. The authors, who have come together to produce this monograph, are all expert in the practical observation of structures using a variety of techniques and technology and the text represents their experiences and knowledge in this important area of structural engineering.

The cost of monitoring a structure will always require sound justification, but there is little doubt that it will often ensure public safety and the best use of resources either by justifying the continued use of a structure or by indicating the necessity to demolish and rebuild.

This text has been prepared for the practising engineer who is required by his client to advise on the safety and use of a building and to provide the most cost-effective solution for perceived problems associated with that building. It will also be helpful to building owners by showing what can be done with established monitoring techniques.

I am grateful to the authors who have all worked hard to produce their contributions which include many practical examples of the application of the techniques which are their particular specialities.

G. S. T. Armer
GA Consultants
Boston, UK
January 2001

Chapter 1

Introduction

G.S.T. Armer

1.1 Failure in the built environment

For most advanced economies, buildings and other construction represent one of the largest, if not the largest, single investment of national resources. The well-being of such economies is heavily dependent upon the satisfactory performance of its construction since this involves not only the provision of protective envelopes for a myriad of economic activities, it also is critical for the infrastructure and the variety of communication systems providing the lifeblood for such activities. Failures in the built environment occur for a variety of reasons and on a variety of scales. In macroeconomic terms, the small domestic structure which burns to the ground because of a fire started in play by children is insignificant when compared with a national disaster, such as a major earthquake, affecting a million people simultaneously. There are also less obvious types of failure which can cause widespread human distress and be of considerable economic significance, such as problems occurring in populations of system-built construction where many thousands of units have to be considered as unsatisfactory as a result of a few actual failures. Examples of this phenomenon of population failure have occurred in the UK in large-panel system-built housing (BRE, 1985, 1986, 1987) and in long-span school and leisure halls (Bate, 1974, 1984; DoE, 1974) amongst others.

In the more temperate climates, failure of the infrastructure usually causes relatively short-term, if expensive, problems. For example, the storms in 1987 and January 1990, in the UK (Figure 1.1) caused disruption of road, rail, air and sea transport, the loss of electricity supply to around a million homes and overloading of the telephone system. For a short time following the latter storm, all the London railway termini were closed. More recently, the flooding problems in the autumn of 2000 have affected many tens of thousands of people, some two or even three times in rapid succession. These types of failure are most commonly associated with what may be called severe natural environmental conditions. It is unusual to

Figure 1.1 Floods in Ottery St Mary.

experience widespread structural failure in these situations although, of course, there are often isolated cases which do occur.

In many areas of the world, such as southern Europe, parts of Australasia and the Americas, the environment can be much harsher with the potential for earthquakes, hurricanes or typhoons to cause substantial destruction with consequent high costs, long repair or replacement times, and great costs in terms of human distress (Figure 1.2).

The design of structural systems to resist the forces generated in each of these extreme load conditions differs between the two cases. However, in both instances the designer is expected to create a built environment which achieves a level of performance acceptable to the public. That is, acceptable in terms of the perceived severity of the hazard. In the event of a major earthquake or typhoon, when the failure of domestic and commercial buildings would not be surprising, then essential construction such as hospitals, power stations and emergency services accommodation would be expected to survive.

1.2 Structural design philosophy

The design options available to the structural engineer to meet his clients' and the public's requirements for buildings which perform satisfactorily within various hazard scenarios may be categorized (Armer, 1988) quite simply thus:

Figure 1.2 Severe earthquake damage to bridge.

(1) Provide a strong basic structure
(2) Provide redundancy within the basic structure
(3) Provide sacrificial defence for the structure
(4) Provide monitoring systems which warn of inadequate structural conditions

The first three of these are relatively familiar ground for the designer.

1.2.1 Structural strength

The principle of making load-bearing structures strong enough to withstand all the loads which are likely to impinge during their lifetime is central to the universally applied structural design philosophy which has developed since the eighteenth century. It provides a straightforward basis on which to establish the level of structural safety by admitting the use of factors attached to either the design loads or to the response, i.e. the strength descriptor. By so doing, the designer is given a direct indication of the margins of safety against failure under the loads he has explicitly identified. The success of this approach depends heavily upon the engineer's skills in predicting both the possible ranges and the combinations of loads which the structure will have to carry and the material and structural performance of the building itself.

1.2.2 Structural redundancy

The principle of explicitly incorporating redundancy into a structure has more recent antecedents. In the early days of aircraft design, there was a major problem with the provision of sufficient power to get the machines off the ground. The two obvious strategies to surmount this difficulty were to build more powerful engines and/or to reduce the weight of the aeroplanes. The latter approach led to some very elegant mathematical solutions for the problem of 'minimum weight design', which in turn led to brittle behaviour of the resultant airframe structures! To overcome this particular difficulty, the concept of 'fail-safe' design was developed, and the idea of beneficial redundancy was introduced into the designer's vocabulary. By the intelligent use of redundant elements, it is possible to ensure that the failure of a single element does not precipitate the failure of the complete structure. As yet, there are no satisfactory theories to guide the designer in this field, but experience and empiricism must suffice.

1.2.3 Structural defence

The provision of sacrificial defence for a building structure is even less scientifically founded. The use of bollards and crash barriers to prevent vehicle impacts is commonplace and their efficacy proven in practice. The use of venting elements such as windows and weak structural elements to reduce the level of explosive pressure on major structural elements is less widely adopted. Weighted trap-doors are sometimes used in factory buildings where severe dust cloud explosions are a high risk, but are arguably not sacrificial. Mainstone (1971) has given considerable data for the design of explosion venting windows, but great care must obviously be taken when using this design strategy. The provision of in-service structural monitoring systems in building construction has so far been limited to a few isolated cases such as the commercial fair complex in the USA (IABSE, 1987). Some systems have found application in civil engineering construction, for example in special bridges and nuclear power stations. It is probably reasonable to assume that the use of monitoring in these circumstances reflects a 'belt and braces' approach to safety rather than a planned use of the technique in an integrated design philosophy.

1.3 The role of structural monitoring

A dictionary definition for 'monitor' is: 'to watch or listen to (something) carefully over a certain period of time for a special purpose'. The end of this definition is perhaps the most important part since it identifies the need to establish purpose as an essential element of the activity of monitoring. There have been a number of examples where considerable quantities of data have been collected, particularly in the field of research, in the vain hope that some brilliant thought or strategy will arise as to what should be done with this information. Many who have been associated with structural testing will be familiar with the thesis that putting on a few more gauges would be 'useful' since more information must be helpful. Unfortunately, insight rarely arrives in such circumstances and, if it does, the inevitable conclusion is that something else should have been measured! So before any consideration is given to methods of observing the performance of construction, the question must be asked: 'Why is this expensive process to be established?' The answer to this question necessarily involves clear understanding of the positive actions which could result from the data-gathering exercise.

The role of construction monitoring has to be established against a background of requirements emanating from the public (i.e. the state), the owner and the user. These requirements have to be expressed in a form which is compatible with the data generated by any monitoring systems which are established. Putting this point more explicitly, whilst it is relatively easy to gather enormous amounts of data from most monitoring systems, it is usually very difficult to interpret such data and to develop useful consequential responses.

Public concern with the performance of building structures is principally directed at those aspects which impinge on the individual, that is upon his own safety and welfare. The acceptable degree of safety is encapsulated in the various building regulations, standards and codes of practice. Unfortunately, there is not a simple correlation between the safety of people in and around construction and the security/safety of that construction, as shown by Armer (1988). Neither is it possible to express a level of safety for a construction as a number. In spite of the large number of academic papers which purport to discuss safety levels in explicit terms and to offer the designer the opportunity to choose his own value, the concept is entirely subjective and acceptability will vary with construction type and location, current political situation and so on.

Any post-construction monitoring of building works carried out by government or state authorities acting as the public executive rather than the owner is essentially limited to the collection of statistical data on failures. This form of monitoring matches the nature of the response which can be implemented practically by a national regulating body. For example, certain types of construction can be outlawed or special requirements can be included in regulations, such as the provision for buildings of over four storeys to be designed to resist disproportionate collapse following accidental damage. The effect of this form of monitoring and response is the control of the performance of the population of constructions and not that of an individual building. This point is sometimes misunderstood in discussions on the function of codes of practice for good design. By giving rules for the design of construction, i.e. codes of practice, the regulating body can ensure that a particular building is part of the population of buildings modelled by a particular code, but it does not, however, guarantee further its quality. Since within any population of artefacts the performance of individual elements will vary from good through average to bad, such will be the lot of construction. Thus the objective of monitoring construction by public authorities is two-fold: first, to ensure that acceptable levels of safety are sustained for the people and, second, to ensure that a degree of consumer protection is provided to meet the same end.

The owner of a building, if not also the user, will require to protect his investment both as capital and as a generator of funds. This can be achieved by regular monitoring and by consequential appropriate maintenance. There are, of course, many quirks in the financial world which may make this simplistic description somewhat inaccurate for particular owners and at particular times but nevertheless it represents a realistic generality.

The user will be concerned that the building he rents or leases provides a safe working environment for both his staff and his business. It is therefore in the user's interest to monitor the building he occupies to ensure that his business is not damaged by the loss of facility.

It is therefore of concern at many levels to ensure the proper functioning of the building stock, and a necessary part of this process is to monitor current condition.

The discussion so far is predicated on the thesis that monitoring construction should be a legitimate weapon in the armoury of those concerned with the in-service life of buildings. It must be admitted, however, that its use in the roles described is only in its infancy. Most practical experience in the instrumentation of building has been gained by researchers. Their objectives have usually been to validate, or sometimes to calibrate, theoretical models of structural behaviour. By so doing they hope to provide another design aid for the engineer. In the technical literature, there are many reported examples of this use of the technique. Ellis and Littler (1988) on the dynamics of tall buildings, Sanada *et al.* (1982) on chimneys, and Jolly and Moy (1987) on a factory building illustrate the variety of work undertaken so far.

Since there is quite a limited number of structural performance indicators which can be monitored, and likewise a limited number of suitable (which is really a euphemism for stable) instruments, there is considerable scope for the exchange of techniques and methods between the various applications for which some experience exists.

References

Armer, G.S.T. (1988) 'Structural safety: some problems of achievement and control', in Beedle, Lynn S. (ed.) *The Second Century of the Skyscraper*, New York: Van Nostrand Reinhold, pp. 741–57.

Bate, S.C.C. (1974) 'Report on the failure of roof beams at Sir John Cass's Foundation and Red Coat Church of England Secondary School', *BRE Current Paper CP58/74*, Watford: Building Research Establishment.

Bate, S.C.C. (1984) 'High alumina cement concrete in existing building superstructures', *BRE Report*. London: HMSO.

BRE (1985) 'The structure of Ronan Point and other Taylor Woodrow-Anglian buildings', *BRE Report BR 63*. Watford: Building Research Establishment.

BRE (1986) 'Large panel system dwellings: preliminary information on ownership and condition', *BRE Report BR 74*. Watford: Building Research Establishment.

BRE (1987) 'The structural adequacy and durability of large panel system buildings', *BRE Report BR 107*. Watford: Building Research Establishment.

DoE (1974) 'Collapse of roof beams', *Circular Letter BRA1108612* (May 1974). London: Department of the Environment.

Ellis, B.R. and Littler, J.D. (1988) 'Dynamic response of nine similar tower blocks', *Journal of Wind Engineering and Industrial Aerodynamics,* 28: 339–49.

IABSE (1987) 'Monitoring of large scale structures and assessment of their safety', *Proceedings of Conference Bergamo, Italy.*

Jolly, C.K. and Moy, S.S.J. (1987) 'Instrumentation and Monitoring of a Factory Building', in Garas, F.K., Clarke, J.L. and Armer, G.S.T. (eds) *Structural Assessment.* London: Butterworths, pp. 134–42.

Mainstone, R.J. (1971) 'The breakage of glass windows by gas explosions', *BRE Current Paper CP26/71*. Watford: Building Research Establishment.

Sanada S., Nakamura, E.A. and Yoshida, M. (1982) 'Full-scale measurement of excitation and wind force on a 200 m concrete chimney', *Proceedings of the 7th Symposium on Wind Engineering, Tokyo.*

Chapter 2

Dynamic monitoring

B.R. Ellis

2.1 Introduction

The term monitoring usually relates to a continual record of performance, but there are many occasions when discrete measurements or surveys are appropriate in the assessment of structures. Discrete measurements often have an advantage over long-term monitoring due to the costs involved and some of the technical problems which are avoided. However, it is important to select the appropriate system to provide the information required by the investigation. Most dynamic investigations related to the assessment of structures fall into the category of discrete measurements. There are occasions when long-term monitoring is required to check either the vibration levels to which the structure is exposed (e.g. vibrations caused by nearby excavations), or to monitor structural response (e.g. building response to wind loading).

This chapter considers a range of dynamic measurements, the principles of which can be applied to almost any structure. For example a key dynamic parameter of a system is its fundamental natural frequency. A plucked violin string will vibrate at its natural frequency, a building will vibrate when the wind blows and even the whole world vibrates when a large earthquake occurs. In each of these situations, the fundamental natural frequency can be measured. However, for the purposes of this chapter the range of measurements will be restricted to civil engineering structures and the examples provided relate to just two types of structure, one floor and several buildings, so that the various measurements can be compared.

Before considering the detail of the measurements it is important to consider the objective of the work and the actual physical characteristics that can be measured. Every structure will have a series of modes of vibration, which are termed normal modes (normal in the mathematical sense). For civil engineering structures it is usually the fundamental mode (i.e. the lowest frequency mode) which will be most important. Each mode will be defined by four parameters namely, natural frequency (f), stiffness (k), mode shape (ϕ) and damping (ζ), although these parameters may vary with the amplitude of motion of the system. The easiest parameter to measure is the fundamental natural frequency and on many occasions this may be the sole objective of the tests, e.g. checking a dance floor to see whether its frequency meets a threshold value above which the dynamic response need not be evaluated. The damping and mode shape are more difficult to measure, and measurements of stiffness require special test equipment. However, all of these parameters are required to determine how a structure will respond to a dynamic load.

Dynamic measurements do have a wider usage than just for ascertaining structural response to dynamic loads. For example, dynamic characteristics can be compared with the equivalent characteristics calculated using a numerical model, and thereby provide feedback to improve the model. In this situation many modes may be used, and if the model is that used for designing a structure, then it is important to be confident that it is accurate. One example of the use of dynamic measurements is with the situation mentioned earlier, where the natural frequencies of the whole world are measured (the fundamental period being 53.9 min or a frequency of 0.0003092 Hz). These measurements are compared with calculations to study the physical composition of the earth, a subject termed terrestrial spectroscopy.

One final application of dynamic measurements is their use for identifying damage in structures, a topic often called integrity monitoring or system identification. Whilst this initially appears to be a good idea there are many difficulties, some of which will be considered at the end of the chapter, and successful applications in civil engineering are few and far between.

2.2 What can be measured

One of the first equations that will be encountered in texts on structural dynamics is the equation describing the single degree-of-freedom visco elastic model, that is:

$$m\ddot{x} + c\dot{x} + kx = p(t) \tag{1}$$

where m is the mass; c is the damping; k is the stiffness; $p(t)$ is the time varying force; $\ddot{x}$ is the acceleration; $\dot{x}$ is the velocity; and x is the displacement.

When the behaviour of buildings or building elements is examined, it is found that it can be characterised by the combination of a number of independent modes of vibration each of which can be described by an equation similar to (1). Moreover, for a number of applications, only the fundamental mode of vibration needs to be considered, hence the simple one degree-of-freedom model can be used to describe how the structure behaves.

It should be noted that when equation (1) is used to describe a mode of vibration, the parameters are all modal parameters and the co-ordinates are modal co-ordinates, not the conventional terms used in static analysis and Cartesian geometry. However, the two are mathematically related.

The inertial force, $m\ddot{x}$, and the stiffness force, kx, are usually much larger than the damping force $c\dot{x}$. The inertial force and the stiffness force act in opposite directions, and there will be one frequency where they will have equal magnitude. At this frequency the inertial and stiffness forces cancel and the response is controlled by the smaller damping force. This frequency is the natural frequency of the system. At frequencies lower that the natural frequency the stiffness force dominates, whereas for frequencies higher that the natural frequencies the inertia force dominates.

The natural frequency can be determined by consideration of just inertia and stiffness forces and this leads to the following equation:

$$\omega = \sqrt{(k/m)}$$

where ω is natural circular frequency, this is related to the natural frequency f and its inverse, the period T by

$$\omega/2\pi = f = 1/T$$

The damping forces play a dominant role resisting forces at the natural frequency of the system (i.e. at resonance) and hence have a key role in many dynamic evaluations. The equation that relates the damping coefficient c to the damping ratio ζ is

$$C = 2\zeta\omega m$$

commonly used in measurements. Substituting these expressions into equation 1 and dividing by m gives

$$\ddot{x} + 2\zeta\omega\dot{x} + \omega^2 x = p(t)/m \tag{2}$$

To describe the terms graphically and explain what they mean, it is worthwhile repeating a diagram shown in most texts on structural dynamics which is the response of a single degree-of-freedom model to a constant force over a range of frequencies (see Figure 2.1).

The curve is defined by three parameters: (1) The natural (or resonance) frequency which for systems with low damping (i.e. most practical systems) can be considered to be the frequency of the maximum response. (2) The

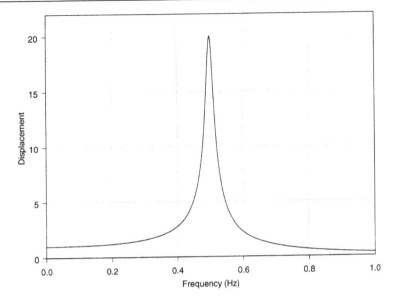

Figure 2.1 The response of a single degree-of-freedom model to a constant force over a range of frequencies.

stiffness, which relates the static force to static displacement (i.e. displacement at zero frequency). (3) The damping which defines the shape of the curve, the lower the damping, the narrower the resonance curve and the higher the amplitude at resonance.

The curve given in the Figure has been generated for a system with 2.5% damping or $\zeta = 0.025$, a stiffness of I and a frequency of 0.5 Hz, and is given to show a typical value which might be found in buildings. A term that is often encountered is magnification factor which relates the response at resonance to the static response and is equal to $1/2\ \zeta$. For this example, the magnification factor is 20.

The other modal parameter that needs to be considered is the mode shape. This is the deflected shape of the structure when it is vibrating in one mode. For the fundamental mode of a simple supported beam this would be a half sine wave, which is not too dissimilar to the deflected shape of the beam under its own weight.

Only one mode has been considered, but there will be the whole family of modes of vibration, albeit each mode can be described by equations 1 or 2. For some structures, modes will also relate to the principal axes of the system. For example, in a tall rectangular building, modes of vibration will be found in both principal horizontal axes of the building, as well as for torsional motion.

The following section deals with the measurement of frequency, damping, mode shape stiffness, thus providing experimental measurements to define the structural system.

2.3 Measurements

Having introduced the principal terms for a dynamic system, it is now appropriate to consider how they can be measured. This is best explained using a series of examples and measurements on two types of structure will be considered. First a composite floor and second several tall buildings. A key part of the measurements will be the equipment used to measure structural response and this is considered first. The various measurements that can be made are then considered starting with the simpler measurements.

2.3.1 Data acquisition and determining frequency

All of the tests require measurements of the structural response to some form of excitation. The measurements of floor vibration will usually require a transducer to be placed at the centre of the floor to monitor vertical motion, and for the output from that transducer to be displayed and/or recorded. The transducer may be an accelerometer, a geophone (or velocity transducer) or a displacement transducer. The simplest form of display is an oscilloscope, and this may be sufficient if only the floor frequency is to be determined, but normally the data will be recorded through an analogue/digital converter and stored on a computer, or spectrum analyzer. Better quality results can be obtained if the signal can be amplified and filtered before recording as this will increase the signal to noise ratio and make better use of the digitisation range of the recording device.

A schematic of the recording system is shown in Figure 2.2. A typical

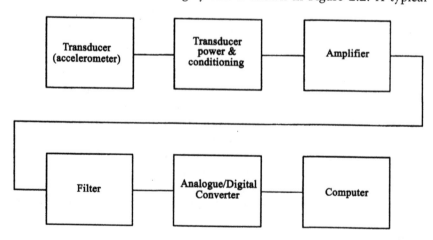

Figure 2.2 A schematic of the recording system.

floor response is given in Figure 2.3. In this case a Shaevitz 2 g servo-accelerometer was used, which has the advantages that it can be used vertically (i.e. with the 1 g offset due to gravity), it is very sensitive and has a large dynamic range down to 0 Hz (cycles/s). Alternatively a 'vertical' geophone would be adequate for most tests.

The signal from the accelerometer was amplified by a factor of 10 and filtered at 30 Hz low-pass to obtain a good quality signal and the data were sampled at 2000 Hz with 8192 data points being recorded. This gives a good resolution for the acceleration time history, but a lower digitisation rate and number of sampled points would be adequate in many cases. From Figure 2.3 which shows the floor motion in a well-defined fundamental mode, the frequency can be derived reasonably easily, i.e. ten cycles take 1.17 s, hence the frequency is 10/1.17 Hz (= 8.55 Hz). However, the usual analysis would involve converting the acceleration–time history into some form of spectrum, relating acceleration to frequency, which is advantageous if more than one mode is excited. The normal means of performing this co-ordinate transformation is using an FFT (fast Fourier transform) procedure to produce an autospectrum. Part of the autospectrum of the data shown in Figure 2.3 is given in Figure 2.4, and the frequency of the mode can easily be identified. The FFT procedure works efficiently with highly composite numbers of points, i.e. numbers that are some 13 power of 2; hence 8192 was selected as being 2^{13}. It is beyond the scope of this chapter to provide the details of such analysis but they are readily available in spectrum analysers or in signal processing software for computers.

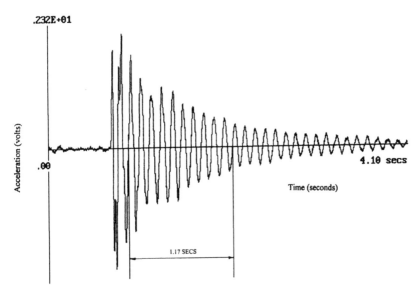

Figure 2.3 A typical measured floor response to an impact.

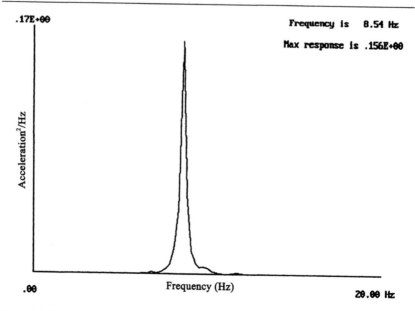

Figure 2.4 Part of the autospectrum of the data shown in Figure 2.3.

Although a variety of transducers has been mentioned, a range of lasers is available which can be used to measure both velocity and displacement. These have the advantage that they do not need to be fixed to the structure, and indeed one system used by the author can take remote measurements without the need for a target to be attached to the structure. Although lasers can be used to provide high quality measurements they have the disadvantage that they are usually expensive.

2.3.2 Measurement of damping

Many test schemes will also require the evaluation of damping, and this is best obtained from a decay of vibration like that shown in Figure 2.5. The simplest method is to measure the amplitude of 'n' successive peaks in the decay (say A_0 to A_n, A) and use the following formula to determine the damping.

$$\zeta = 1/2n\pi.\log_e A_0/A_n \tag{3}$$

This example shows how damping can be estimated from an experiment where a single mode of vibration is encountered with relatively low damping. If the measurements do not resemble a simply decaying sinusoid, then this method of evaluation should not be used. Often, a more complex decay will be obtained, in which several modes of vibration are combined. Given

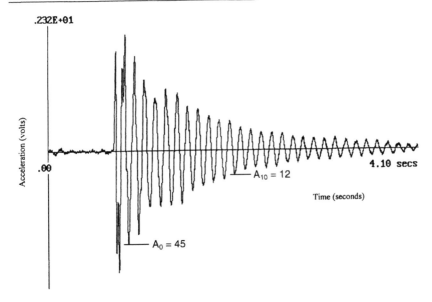

.232E+01

Acceleration (volts)

.00

$A_0 = 45$

$A_{10} = 12$

4.10 secs

Time (seconds)

Figure 2.5 Evaluating damping from a decay of vibration. $\zeta = \dfrac{1}{20\pi} \log_e \dfrac{12}{45} = 2.1\%$

computing facilities it may be possible to filter out the higher modes to leave a good decay and damping can again be estimated, although experience suggests that this is likely to produce slight overestimates.

Another method of estimating damping is from a spectrum, like the one shown in Figure 2.4, and such methods are readily available on spectrum analysers. However, for the correct use of such procedures, it is essential to have a well-resolved spectrum and consider the statistical errors involved, otherwise damping can be overestimated greatly.

Figure 2.1 shows a spectrum of the response of a single degree-of-freedom system to white noise (i.e. constant force at all frequencies). This curve is defined by three parameters, stiffness, natural frequency and damping. If the curve can be derived experimentally, the damping can be obtained.

The damping defines the shape of the curve and amplification at resonance. The usual way of estimating damping is termed the half power bandwidth method and is shown in Figure 2.5. It involves measuring the response at resonance R and determining the frequencies f_1 and f_2 at which the response is $R/\sqrt{2}$. These frequencies are then used in the equation to determine the damping.

$$\zeta = (f_2 - f_1)/(f_2 + f_1)$$

More complex algorithms can be used to make use of all the data points in the curve to provide a better estimation of the damping using a best-fit, viscoelastic model.

2.3.3 Testing methods and procedures for floors

In this section, five different methods for testing floors will be considered, although the first and third are primarily given for illustration purposes. For each test the test procedure is discussed, and it is assumed that a transducer is set up to record floor motion. A summary is given in Table 2.1.

2.3.3.1 Ambient

All structures vibrate in response to naturally occurring excitations, e.g. air movement. Thus naturally occurring floor motion can be monitored using the recording system, and as the floor will respond principally in its normal modes of vibration, these can be identified by examining the spectrum derived from the recorded response. This is not a method that would normally be used as it requires a relatively long recording period and a sensitive monitoring system. However, it has the advantage that it can be used on any system, however large.

On the test floor, the response was recorded in twelve files each of just over 40 s and the accelerometer output had to be amplified by a factor of 10,000. The spectra from each of the twelve files were summed, and as the resulting spectrum was very ragged it was smoothed to produce the spectrum shown in Figure 2.6. It can be seen that the main mode is clearly identified (8.59 Hz), and several other modes can be seen. The lower frequency modes shown in this spectrum relate to other areas of the floor which were loaded with sandbags.

If the damping is determined for the 8.59 Hz mode shown in the spectrum, using a curve-fitting technique, a value of 4.1% is obtained, which is a considerable overestimate. This is not recommended as a sensible approach for floors.

Table 2.1 Summary of test methods

Test	Freq.	Damp	Stiffness	Mode shape	Comment
Ambient	●	×	×	×	Not reasonable for floors
Heel drop	●	●	×	×	The simplest test
Sand bag	●	●	×	×	Straightforward
Impact hammer	●	●	●	?possible	Towards specialist
Forced vibration	●	●	●	●	The most detailed

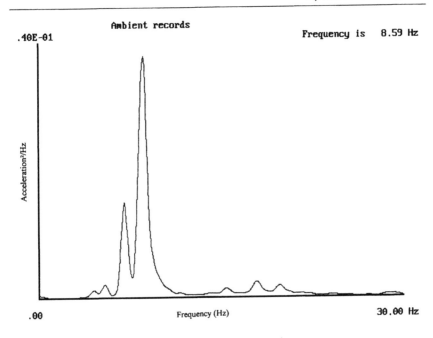

Figure 2.6 Smoothed autospectrum from ambient records.

2.3.3.2 Heel drop test

This is one of the most widely used tests and is very simple. A person stands near to the centre of the floor, then when requested raises himself on to the balls of his feet and suddenly drops on to the heels of his feet, thus creating an impact that is monitored. A typical record is given in Figure 2.7. As shown earlier, both frequency (8.53 Hz) and damping (2.66% crit can be derived from such a record. A potential drawback of the method is that the tester is standing on the floor and will thus affect its characteristics.

2.3.3.3 Sand bag test

This is not a test generally used, but again serves to illustrate a point. It is similar in principle to the heel drop test, because it provides a single impact, but avoids having a person standing on the floor. The test is simply to toss a bag containing 1 kg of loose sand on to the floor to land with a vertical trajectory near to the centre of the floor. There is of course no significance to the actual weight of sand, and it would be possible to think of dropping a given weight through a known height to try to calibrate the method and interpret the magnitude of the floor response. The response to a simple sand bag impact is shown in Figure 2.2, which has a frequency of 8.60 Hz and damping of 1.81% crit (from curve fitting).

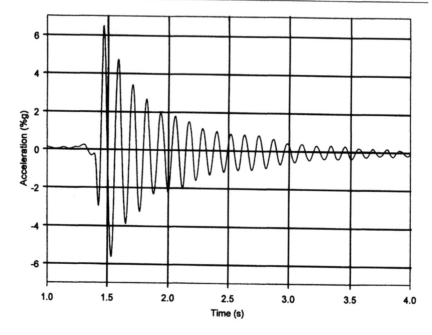

Figure 2.7 Response to a typical heel-drop.

To show the effects of a person on the floor, i.e. human–structure inter-action, the test was repeated but with a person standing on the floor and simply dropping the sandbag from a height of I in. This gave a frequency of 8.62 Hz and damping of 2.37% crit, i.e. the presence of the person increased the damping.

2.3.3.4 Impact hammer

With this test, the load is monitored and this leads to the important possi-bility of evaluating stiffness. An impact hammer for this type of test, is simply an appropriately sized hammer with a compliant head which con-tains a load transducer. The compliant head is to soften the impact in an attempt to introduce force over a range of frequencies, of which the lower frequency end is all important. The output from the load transducer is recorded along with the floor response and with similar recording para-meters, when the floor is impacted by the hammer. With this test a person will be standing on the floor, hence that person will interact with the floor and affect its frequency and damping.

Impact hammer tests are used extensively in mechanical engineering, but are less common in civil engineering where the larger lower frequency struc-tures present a variety of problems. Perhaps the most significant problem is because the impact is short in the time domain to give a broad spectrum in the frequency domain, yet the response is long in the time domain but very narrow in the frequency domain. Thus the data are far from ideal for analysis. The testing procedure is to excite the floor by means of the hammer impulse, and to measure the force and resulting vibrations. The force and response are transformed into the frequency domain from which the frequency response function (FRF) is derived. The modal parameters can then be extracted from the FRF using curve fitting. The FRF obtained is shown in Figure 2.8 and is not considered to be sufficiently good to extract meaningful data. The test was repeated twelve times using a small hammer with the softest head, but the system is better for testing systems with higher frequencies.

The main advantage of the hammer tests is that they can be undertaken quickly, and if a spectrum analyser is available, the analysis can be performed almost instantly. The problem is that this system will provide answers whatever the quality of the data and hence the operator needs to be careful to provide his own quality checks. When this form of testing has been used on civil engineering structures, often only the frequency and damping are determined and this reduces their value to that of the heel-drop test described earlier. Also if the analysis is conducted in the frequency domain

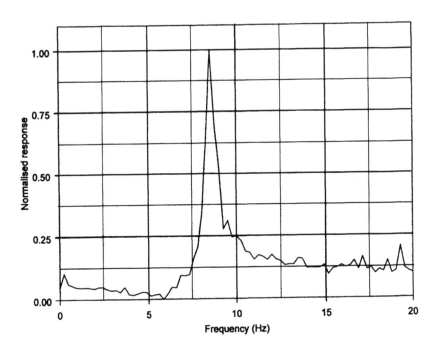

Figure 2.8 A FRF from a hammer test.

then the damping estimates will be much worse that those obtained by analysis of decays in the time domain.

With this scheme it would be possible to take mode shape measurements by using two transducers, one reference and one which can be placed at selected positions. For each impact both responses could be measured and the ratio of responses taken, either via a frequency domain method that would suffer from the potential lack of resolution of the FRF, or through a time domain approach comparing the two signals. Either way this would be a relatively slow and not very accurate process.

2.3.3.5 Forced vibration testing

To measure the stiffness and mode shape of a mode of vibration, a comprehensive test set-up is required. This involves the use of a vibration generator to vibrate the floor in a controlled manner. The vibration generators used by the author are simple in concept and use a pair of contra-rotating masses to provide a unidirectional sinusoidal force. Only a very small generator is required to test a floor, and the small generators used by the author are all powered by stepper motors that have a frequency resolution of 1600 steps/Hz.

The test scheme is basically the same whether the test specimen is a wall, a floor or a complete building, albeit different sizes of vibration generator are required. The general test scheme is as follows:

(1) The vibration generator is rigidly attached to the structure at a position where the maximum modal force is likely to be exerted (i.e. at the centre of a floor).
(2) A transducer is placed near the vibration generator to monitor motion in the desired direction. The output from the transducer is amplified (if necessary) and filtered before being displayed on an oscilloscope and fed into a computer.
(3) The generator is adjusted to produce the desired force level.
(4) The frequency of excitation is incremented through a selected range and the response recorded at each stage.
(5) The plot of frequency *versus* response is normalized by converting the measured motion to equivalent displacements and then dividing the displacement by the applied force. The plot of normalised response against excitation frequency is herein termed the response spectrum.
(6) The natural frequencies are identified by peaks in the spectrum. The actual values of the natural frequency, damping and stiffness for each mode are determined by fitting a multidegree-of-freedom model to the response spectrum.
(7) A further measurement of damping for each mode is obtained by suddenly stopping the excitation at the natural frequency and recording the resulting decay of oscillation.

(8) The mode shape for each mode is obtained when the structure is sub-jected to steady-state excitation at its natural frequency, by moving a second transducer to various positions on the structure and comparing the response with that measured using the reference transducer.
(9) The whole procedure is repeated to monitor modes in other directions and also to examine the effects of varying the force generated.

The response spectrum obtained for the tests on the floor is shown in Figure 2.9. The crosses on the Figure represent the measured values and the continuous line represents the best-fit single degree-of-freedom curve. The values of frequency, damping and stiffness are the parameters that define this curve and hence provide the best estimate of the equivalent para-meters defining the modes of the structure. The analysis of the decay pro-vides an independent measurement of the damping parameter, although it is often only useful for the fundamental mode. For the example presented, the frequency and the damping from the decay measurements were 8.48 Hz and 1.64% (Figure 2.10) and the curve fitting gave 8.49 Hz and 1.65%, which gives an indication of the accuracy of the measurements.

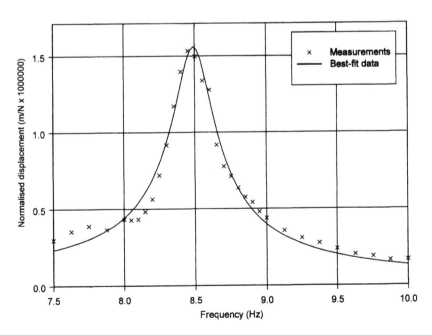

Figure 2.9 The response spectrum obtained for the tests on the floor.

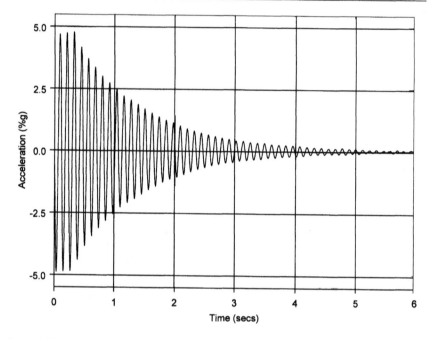

Figure 2.10 Decay measured during a forced vibration test.

2.3.3.6 Comparative results

The various tests have been discussed earlier. Table 2.2 is provided simply for ease of comparison of all of the test results.

The slight variation in frequency is due to two factors. First, the fact that the tests have different initial amplitudes of vibration which affects the frequency, and second for several of the tests a person was standing on the floor. This second factor has a much larger influence on the measured damping.

Table 2.2 Summary of results from all tests

Test	Frequency (Hz)	Damping (% crit)	Stiffness N/m
Ambient	8.59		
Sand bag	8.60	1.81	
Sand bag + person	8.62	2.37	
Heel drop	8.53	2.66	
FRF (Hammer)	8.54	—	—
FVT (curve fitting)	8.49	1.65	1.94×10^7
FVT (decay)	8.48	1.64	

2.4 Testing buildings

In principle the range of tests described earlier can be used on many different types of structures, but very large structures can present different problems, and for illustration purposes consider tall buildings. Here the simple
'impact' tests using the sandbag, heel drop or impulse hammer are not
appropriate, albeit a variety of experiments have been undertaken where
an impulse has been applied to a building. However, this type of impulse
test is not recommended.

Perhaps the simplest way to measure the natural frequency of a structure
subjected to wind excitation is to record the signal from a measurement
transducer and, if the signal resembles a single sine wave, determine the
time between successive peaks in the time history. This gives the period of
the motion that is the inverse of the frequency. This method works well
when the response is primarily in one mode. Careful positioning of the
transducer and possible filtering of the signal may be required in order to
minimise interference between translational and torsional modes.

Another fairly simple way of measuring frequency is to make a short
recording of the response of the structure to wind loading (a response–time
history). The short recording could be of several minutes duration. The
record can then be transformed from the time to the frequency domain
using a fast Fourier transform (FFT) routine to produce an autospectrum.
If measurements of structural characteristics other than frequency are
required then there are two main alternatives, forced vibration tests and
spectral analysis of ambient motion. The former is required if stiffness is
to be measured. The latter may be the only option for very large structures
like suspension bridges.

2.4.1 Forced vibration tests and observation of nonlinear characteristics

The forced vibration method that was described for testing floors can be
used for testing buildings. However, the vibration generator will need to
be far larger. The large vibration generators used by the author have a
frequency range of 0.3–20 Hz and can produce a maximum peak-peak
force of 8.2 tonne. The force can be selected within a relatively wide range
and this enables the nonlinear characteristics of a structure to be examined.
The system consists of four individual force generators and has been used
to test a range of large structures the largest of which was a 220 m tall
arch dam. This system can only be used to provide horizontal excitation.
Each generator requires four people to lift and manoeuvre it into position.

It has been mentioned that structures do exhibit certain nonlinear characteristics and it is worthwhile considering what is meant by a linear material
or a linear system. A material is considered to be linear if its Young's modu-

lus is constant, i.e. if stress is directly proportional to strain. However, if the material is damped, the plot of stress against strain would exhibit a hysteresis loop, or part thereof, so that no real material is truly linear. The behaviour of a system can often be described by a differential equation, and the system is called linear if the influence coefficients related to the differential terms are constant. For the behaviour of structure to be termed linear requires both the damping and natural frequency of each mode to be constant. The damping values and natural frequencies which are observed in practice do vary with the amplitude of motion and to demonstrate these nonlinearities it is worthwhile examining the results obtained from tests on two buildings.

Hume Point is a 22-storey large panel building, in East London, and the details of the vibration tests are given in Ellis and Littler, 1988. The frequency sweeps were repeated for a range of forces and visco-elastic curves were fitted to the experimental data to determine the dynamic characteristics. The results are shown in Table 2.3. It can be seen that the resonance frequency gradually decreases and the damping ratio increases with increasing amplitude of motion.

Another means of obtaining information on nonlinear characteristics is through detailed examination of vibration decays recorded during forced vibration tests. To illustrate this method consider an eight-storey steel-framed building. To determine values of frequency and damping a best-fit decay based on a visco-elastic model is derived for a selected part of the measured decay, using a least squares fitting procedure. The fitting procedure varies frequency, damping and initial amplitude. Five contiguous 10-s samples were selected from the measured decay and visco-elastic decays fitted to the measured data. The tail of the measured decay is disregarded as this is affected by the ambient excitation of the structure. The data extracted from the five samples are given in Table 2.4 and related to the amplitude of vibration at the start of the first sample. The change in

Table 2.3 Characteristics of NS I mode of Hume Point

Maximum response measurements				Curve fitting		Decay
Frequency (Hz)	Force (kN)	Acceleration (m/s²)	Displacement (mm)	Frequency (Hz)	Damping (% crit)	Damping (% crit)
1.100	10.01	0.12	2.4	1.104	1.20	1.20
1.100	8.01	0.097	2.0	1.105	1.22	1.16
1.105	6.06	0.074	1.5	1.107	1.11	1.13
1.105	4.04	0.051	1.0	1.110	1.13	1.15
1.110	2.06	0.027	0.55	1.114	1.03	1.15
1.120	1.04	0.015	0.30	1.121	0.97	1.05

Table 2.4 Frequency and damping evaluated for
various sections of a decay

Relative amplitude	Natural frequency (Hz)	Damping (% crit)
1.000	0.611	2.87
0.366	0.636	1.81
0.181	0.645	1.28
0.106	0.647	1.02
0.062	0.656	0.85

frequency that was observed covered the variation observed between the results of ambient and forced vibration tests on the building.

The results illustrate the characteristics that have been seen in many structures for relatively low amplitude motion, that is the frequency increases and the damping decreases with decreasing amplitude of motion. While the characteristic changes can be observed, it can be appreciated that the changes are relatively small and can be ignored for most purposes. It does, however, suggest that there should be slight differences observed in the results of different types of test which do excite different amplitudes of vibration.

2.4.2 Spectral analysis of ambient motion

The forced vibration tests introduced a controlled force to excite the structure and thus provide a deterministic test method. However, the wind also causes structural vibrations, and the analysis of these vibrations can be used to determine structural characteristics. Analysing the wind response of a structure does not require specialist equipment. However, the analysis may prove to be quite a complex task, and the problems that may be encountered are discussed in the following sections.

2.4.2.1 Analysis using relatively short records

In order to examine the response of structures to wind loading, research workers have often used the spectral analysis of relatively short lengths of data, but its limitations are still not always realised. In essence, a recording of the response of the structure to the wind is made for a period of several hours. This recording is then split into a number of individual records and an FFT calculated for each one. The resulting spectra are then averaged together to give an ensemble averaged spectrum from which values of natural frequency, damping and response can be obtained for each mode. There are three important criteria to consider when using this method:

(1) The stationarity of the data
(2) Variance errors
(3) Bias errors

In order to use spectral analysis methods correctly, it is necessary to have stationary data for analysis. Stationary data have statistical properties which do not vary with time. If the wind direction or speed varies over the length of the recording, then the response in any particular mode will vary in amplitude and will not be stationary. Averaging spectra which were obtained under different wind conditions together will not give the same spectrum that would be obtained from stationary data recorded for the average of these wind conditions. Therefore a test for stationarity should be performed on the whole recording. If, as is usually the case, the data are not stationary, then extreme caution should be used in extracting any parameters other than frequency. If damping values are obtained, then they will inevitably overestimate the real values.

Variance or random error results from the fact that any analysis must be performed on a finite number of sample records. On an intuitive level, the more times a random process is sampled, and the results averaged, the greater the accuracy, providing the same process is sampled each time (i.e. it is stationary). Variance error is equal to the reciprocal of the square root of the number of records averaged together. Therefore, for example, variance error is ±10% for 100 records, but only reduces to ±5% when 400 records are used.

Bias errors occur when there is a rapid change of amplitude with respect to frequency and there are too few spectral lines to resolve the variation accurately. Such a situation occurs in the region of a peak in the response spectrum, i.e. at a natural frequency. In order to limit bias errors to 2% there must be at least four spectral lines in the half-power bandwidth of each mode (Bendat and Piersol, 1986).

Despite the problems outlined above, natural frequencies can be obtained quickly and easily by this method and satisfactory mode shapes may be also be obtained providing the response measurements at different positions are made at the same time.

2.4.2.2 Selective ensemble averaging

Selective ensemble averaging tries to overcome the problem of obtaining stationary data of sufficient accuracy by forcing an ensemble to contain stationary data. This is done by making many recordings, or an extremely long continuous recording, and dividing the data into individual blocks, or records, of the length required for analysis, and storing the individual records along with the average wind speed and direction for the length of the record. Records obtained under similar wind conditions are then selected and averaged. This type of analysis was undertaken on data recorded on Hume Point and is described in Littler and Ellis (1992).

Selective ensemble averaging is primarily used for obtaining the response of a structure to a given wind to a high degree of accuracy. The results obtained using this method may be used in a comparison with those obtained from calculation or wind tunnel studies to help calibrate codes or wind tunnels. Although this method yields accurate dynamic characteristics (natural frequency, damping and mode shapes), it is unlikely to be worthwhile employing this technique if it is only these characteristics that are required from the test. The exception to this would be a structure with very low natural frequencies and damping such as a long-span suspension bridge where forced vibration tests or spectral analysis of short ambient recordings is unlikely to be successful.

2.4.2.3 Other methods of analysing wind-induced response

A number of analysis methods have been developed which use much shorter lengths of data than are required by conventional spectral analysis. One of these methods is the autoregressive (AR) model which uses parameters derived by the maximum entropy method (MEM) and this method has been examined using measured data from Hume Point (Cao *et al.*, 1997). Two other related models are the moving average (MA) and the autoregressive moving average (ARMA), which both use parameters derived by the maximum likelihood method (Bendat and Piersol, 1986). With all of these types of analysis the use of stationary data is very important. Whilst all of these methods use much shorter lengths of data than standard spectral analysis, the accuracy obtained is dependent upon the model used. If an inappropriate model is selected, then the results will be much worse than those obtained by conventional analysis. At present, there are no available methods of assessing the likely accuracy of the spectra derived using these methods, and while they may be used for estimating modal frequencies, their use for estimating damping or modal response must be questioned. Nevertheless, the methods have considerable potential, although they may require further refinement and evaluation before they can be used with confidence to evaluate structural characteristics.

2.5 Use of information

So far, the paper has dealt with how to measure various structural parameters and how accurate the measurements are likely to be. This section deals with how the information can be used.

2.5.1 Frequencies

When a measurement of the fundamental frequency of a structure has been made, there are a number of uses to which it can be put; however, they all require another frequency value for comparison. Three comparative values can be used, namely:

(1) The average frequency of similar simple structures
(2) The frequency of the structure measured previously
(3) The calculated frequency of the structure.

Each of these will be discussed.

2.5.1.1 Comparison of measurement with measurements from similar simple structures

Some relatively simple structures or structural elements may occur frequently, and if the integrity of these elements is of concern, then it can be very useful to compare the characteristics of the nominally similar structures. Three examples of such simple structures are:

(1) Decorative pinnacles
(2) Cladding panels
(3) Light standards.

To illustrate this, consider decorative pinnacles. These are found on many historic buildings and the Palace of Westminster in London has over 500 pinnacles that have been monitored (Ellis, 1998). The pinnacles are of various sizes and types, but in the vast majority of cases there are sufficient pinnacles of a given type for comparative analysis. The pinnacles are often positioned in relatively inaccessible places and this is an example where remote measurement using a laser is ideal. Like most structures, pinnacles will deteriorate with time and significant deterioration may cause them to fall, occasionally causing substantial damage to the supporting structure. This is evident in many buildings throughout Europe. The problem is often deterioration along mortar joints, and this mechanism of damage will affect both strength and stiffness and hence can be identified through frequency measurements. Therefore if the frequencies of similar pinnacles are compared, any pinnacle with significantly different characteristics to the average can be identified, with those with significantly lower frequencies possibly indicating damage. The question of when the differences can be classified as significant is difficult to answer in general terms, and this will probably have to be based on experience of repeated measurements and detailed examination which is not available at present. Suffice to say that for some structures the differences are only too obvious, both with regard to frequency and amplitude of vibration.

2.5.1.2 Comparison of measurement with previous measurements on the structure

The comparison of frequency measurements made at various times on a structure has often been proposed as a means of continued integrity assessments, and although it may be a simple process to take such measurements, the correct interpretation of the data may prove to be exceedingly difficult. The simple reasoning behind this type of test may be as follows: the frequency relates stiffness and mass; damage to the structure will affect stiffness, but not mass; hence, damage can be identified by frequency changes. However, consider some of the problems involved.

(1) Although the fundamental (or higher) frequency can be measured quite accurately, it will vary with a number of parameters. Three examples of which are: daily variations due to differential heating from the sun; seasonal variations due to changing temperatures; and variations due to nonlinear characteristics at different excitation levels, i.e. related to wind speed or direction. So, for an undamaged structure, a range of frequencies will be observed.
(2) Damage to a structure may affect some modes of vibration, but not others. For example, damage may occur to a floor, but may not affect the fundamental mode of the building, albeit it would affect the localised floor mode.

Most test methods determine elastic parameters like frequency or stiffness, rather than strength, so some form of link between the two needs to be established. This may be evident for very simple structures like pinnacles, where deterioration across joints may lead to reductions in both strength and stiffness, but may be difficult to establish for more complex structures, like bridges or offshore structures. Therefore, to identify damage using repeated measurements of frequency it is important to understand the damage mechanism, to know what modes to examine, and to be confident that the changes which are of consequence are larger than the resolution in the measurements and the nonlinearities in the structure.

2.5.1.3 Comparison of measurement with calculations

One of the most important aspects of post-construction testing is to provide information to verify that a structure is behaving in a similar manner to that assumed in its design. A measurement of the fundamental frequency of a structure or structural element can be compared with a calculation of the same parameter to provide a check on the accuracy of the calculations. As the frequency can be calculated from the simple relationship between stiffness and mass, it can be appreciated that this is a verification of the model used to design the structure. If the designer cannot work out the stiffness and mass of the structure, then he will not know how the loads are carried by the structure or what stresses are encountered or whether the stresses are acceptable.

It is not always possible to calculate the fundamental frequency of a structure accurately and providing feedback to check calculations is useful for two reasons. First, in assessing accuracy of the assumptions and calculations used in the design of that particular structure and second in learning how accurate mathematical models are likely to be in general. This second point will enable weakness in design assumptions to be identified, it will provide data to enable research to concentrate on real problems and so will lead to a better understanding of structural behaviour.

One problem that the author has encountered on many occasions is that it is often assumed that calculations are far more accurate than they really are. In 1980, the author did a survey of publications (Ellis, 1980) in which both computer-based calculations and measurements of the fundamental frequencies of tall buildings were given. From this survey a sample of data was selected for rectangular plan buildings where calculations were not influenced by measurements. Whilst it was thought that this might be a somewhat biased sample, it turned out that the simple empirical relationships were more accurate that computer-based calculations. With empirical calculations errors of more than ±50% are not uncommon, so it can be seen that calculations of fundamental frequencies may not be too accurate.

2.5.2 Damping

Damping may appear to be of little use on its own. Indeed, if the literature reflected typical (and accurate) damping values for a wide range of structures, then measuring damping by itself would have little meaning.

For cases where resonance may occur, methods of calculation are often given which require a damping parameter. For example, methods to determine the response of floors to vibration require a damping parameter (for the fundamental mode). As this parameter is usually inversely proportional to the calculated response it is quite important.

2.5.3 Mode shapes

Mode shapes are usually measured in forced vibration tests although they can be obtained from ambient response measurements. The mode shape can be useful along with the measured frequency for checking calculations, because accurate calculations should predict the correct frequency and corresponding mode shape. Alternatively if the frequency prediction is incorrect, then the mismatch in mode shape might indicate where the assumed stiffness distribution is incorrect.

One area where difficulties often arise is in modelling the restraint conditions for a particular structure. For example, it is often assumed that the base of a tall building is rigid, however, the measured mode shape might identify movement at the base of the building indicating that soil structure interaction is significant. Alternatively, the problem can be inverted, in that measured mode shapes (along with stiffness and frequency) can be used to evaluate the effect of the restraints.

2.5.4 Stiffness

When a forced vibration test is conducted and the frequency, damping, mode shape and stiffness of each relevant mode of vibration has been measured, then the system is well-defined. These are the parameters required to predict the dynamic (or static) behaviour of a structure within the elastic region. For example these parameters are required in conjunction with the actual wind load, to predict how a building will respond to wind loading, and an accurate knowledge of these parameters is obviously a good basis for the calculation.

References

Bendat, J.S. and Piersol, A.G. (1986) *Random Data Analysis and Measurement Procedures* (2nd edition). New York: John Wiley.

Cao, H., Ellis, B.R. and Littler, J.D. (1997) 'The use of the maximum entropy method for the spectral analysis of wind-induced data recorded on buildings', *Journal for Wind Engineering and Industrial Aerodynamics*, 72: 81–93.

Ellis, B.R. (1980) 'An assessment of the accuracy of predicting the fundamental natural frequencies of buildings and the implications concerning the dynamic analysis of structures', *Proc. Instn Civ. Engrs part 2*, 69: 763–76.

Ellis, B.R. (1998) 'Non-destructive dynamic testing of stone pinnacles on the Palace of Westminster', *Proc. I.C.E. Structs and Bldgs*, 128: 300–7.

Ellis, B.R. and Littler, J.D. (1988) 'Dynamic response of nine similar tower blocks', *Journal of Wind Engineering and Industrial Aerodynamics*, 28: 339–49.

Littler, J.D. and Ellis, B.R. (1992) 'Full-scale measurements to determine the response of Hume Point to wind loading', *Journal of Wind Engineering and Industrial Aerodynamics*, 41–44: 1085–96.

Chapter 3

Photogrammetry – theory and technology

Donald Stirling

3.1 Introduction

Photogrammetry is the science of obtaining reliable spatial information, usually in the form of three-dimensional co-ordinates, by taking measurements of two-dimensional photographs. A rather more tongue-in-cheek description is that photogrammetry is 'the art of avoiding measurement'. Basically any complex three-dimensional object can often be measured and modelled more efficiently by using photogrammetric techniques instead of conventional surveying or manual measurement methods.

The inventor of photogrammetry is generally accepted to be the Frenchman, Aimé Laussedat, who pioneered the use of terrestrial photogrammetry for architectural recording around 1860. Towards the end of the 19th century, terrestrial photogrammetric techniques were applied to alpine and glacier mapping. Around 1900, a number of important developments took place in instrument design, including the production of the first stereocomparator in 1901, and the first analogue stereoplotting instruments for use with terrestrial photographs in 1908. This was a major development as the mathematics of photogrammetry involve many complex calculations which, around 1900, involved long and tedious manual computation. The development of analogue (mechanical) computers to solve these computations greatly improved the speed and reliability of the photogrammetric solution, thus increasing its usage. World War I provided the impetus for the rapid development in aircraft design which in turn caused a shift in the emphasis to aerial photogrammetry using air photographs for reconnaissance and mapping. The first analogue stereoplotting instrument for aerial mapping was developed in 1921 and aerial mapping soon became the dominant application of photogrammetry, the use of terrestrial photogrammetry becoming restricted to specialist applications, particularly in architecture and glaciology. The development of the digital computer enabled analytical photogrammetry, with its many complex calculations, to make a comeback and today analogue photogrammetric instruments are virtually obsolete. The flexibility afforded by the digital computer has enabled photogrammetry to develop into many fields and non-mapping applications of photogrammetry abound including in medicine, geomorphology and many fields of engineering including aerospace, mechanical, automotive, shipbuilding and construction.

3.2 Principles of photogrammetry

Figure 3.1 illustrates the situation that exists when an object is photographed. The camera produces a central perspective projection of the object on the negative where the centre of the camera lens, O, is the perspective centre for the projection. It is normal to consider that all rays of light from an object pass through a single point at the centre of the camera lens assembly. In reality a camera lens is a complex optical assembly comprising many different pieces of glass and light rays from different parts of the object photographed can take different routes through the lens. However, these variations are very small and the assumption of a single perspective centre is a valid one. Therefore a point A on the object is imaged at a′ on the negative and an object point B is imaged at b′. The angle θ subtended at O by A and B is recreated inside the camera by θ′, the angle subtended at O by a′ and U. The mathematics of photogrammetry, in effect, allows any desired angle to be recreated from measurements between two image points on the photograph. By measuring the positions of a series of image points on a photograph the resulting series of angles produces what is known as a bundle of rays.

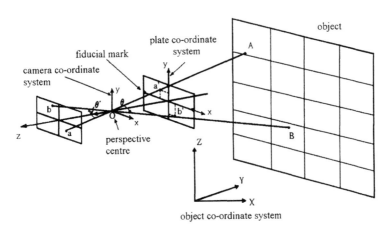

Figure 3.1 Photogrammetric co-ordinate systems and the geometry of a central perspective projection.

In reality the mathematics of photogrammetry is simply a series of transformations between various co-ordinate systems.

3.2.1 Co-ordinate systems

Figure 3.1 also illustrates the three main co-ordinate systems which are used in photogrammetry.

There is a three-dimensional right-angled co-ordinate system for points on the object, X Y Z, known as the object co-ordinate system. This can be any suitable system chosen for convenience for the particular measuring task. It could, for instance, be the Ordnance Survey National Grid where the co-ordinates would be given as Eastings, Northings and Height. However, in monitoring tasks it is more likely that this co-ordinate system will be based either on a local site grid or as a co-ordinate system aligned with a principal axis of the structure which is being monitored. For instance if a building facade was being monitored it would be normal practice to devise a co-ordinate system where, say, the x-axis was aligned with the main front of the building.

The two-dimensional photograph has a rectangular co-ordinate system generally referred to as the photo or plate co-ordinate system, x y. This can either be based on the negative image plane, the situation in the camera when the photograph is exposed, or, more commonly, on the positive image plane which is the way photographs are usually viewed and measured, i.e. the image appears the right way up and the right way round. Thus, from Figure 3.1, point A will appear on the negative image plane at a' and on the positive image plane at a with positive plate co-ordinates x_a, y_a. Similarly, B will appear at b' on the negative image plane and at b on the positive image plane with positive plate co-ordinates x_b, y_b. It will be assumed from here on that all measurements of plate co-ordinates refer to those taken on the positive image plane.

The third co-ordinate system to be considered is the camera co-ordinate system with its origin at the perspective centre, 0. This is another three-dimensional right-angled system. The x and y axes of this system are parallel to the x and y axes of the plate co-ordinate system. The z-axis is in the direction from the perspective centre towards the negative plane. Therefore an image point with plate co-ordinates x_a, y_a has camera co-ordinates of x_a, y_a, z_a. The variable name f, for the focal length of the camera, is often substituted for Z_a.

There are a number of other co-ordinate systems which may be used in some solutions in photogrammetry. When using conventional film cameras it is often not practical to measure plate co-ordinates directly so an additional two dimensional co-ordinate system may be required, the machine or comparator co-ordinate system, with an arbitrary origin and rectangular axes, X_c Y_c. This is a convenient system based on the measuring axes of the instrument.

3.2.2 Rotation systems

A camera is free to move in space along the three axes of the camera co-ordinate system. It can also rotate around these three axes. Thus a camera is said to have six degrees of freedom, three translations and three rotations. The x camera axis is taken as the primary axis of rotation and the rotation around this axis is known as the -rotation which is positive if it is clockwise when viewed in a positive direction along the axis of rotation. The secondary axis of rotation is the y camera axis and the rotation around this axis is known as φ. The z camera axis is the tertiary axis and this rotation is known as κ. These three rotations operate sequentially, i.e. the co-ordinate axes tilt with each applied rotation so that the secondary φ-rotation takes place around a y axis which has already been tilted by the primary ω-rotation around the x-axis. Finally, the tertiary κ-rotation occurs around the doubly rotated z-axis.

3.3 Mathematical models used in photogrammetry

The mathematics of photogrammetry consists of a series of transformations between the various co-ordinate systems detailed above.

3.3.1 Vertical photography

Figure 3.2 illustrates the situation when the camera co-ordinate axes (x y z) are parallel with the corresponding object co-ordinate system axes (X, Y, Z). The perspective centre O, the origin of the camera system, has object co-ordinates of X_O, Y_O, Z_O. (The photograph is shown as a positively viewed image.) A point A on the object, with object co-ordinates X_A, Z_A, is imaged by the perspective centre on the photograph at point a with plate co-ordinates x_a, y_a and corresponding camera co-ordinates x_a, y_a, z_a.

Three projective transformation equations can be written as:

$$X_A - X_O = \lambda_a . x_a \tag{1}$$

$$Y_A - Y_O = \lambda_a . y_a$$

$$Z_A - Z_O = \lambda_a . z_a$$

where

$$\lambda_a = \frac{Z_A - Z_O}{z_a}$$

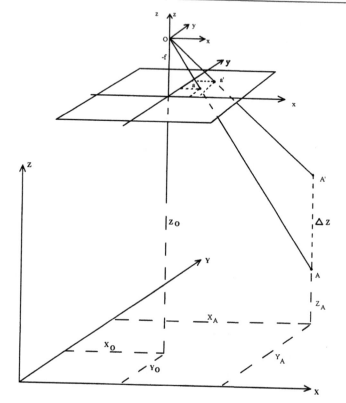

Figure 3.2 Vertical photography with camera and object co-ordinates systems aligned.

λ_a is known as the photo scale factor. Normally when describing the scale of the photograph $1/\lambda_a$ is used and is equivalent to f/D where f *is* the focal length of the camera and D is the taking distance.

A second object point A', a height ΔZ directly above A, is imaged on the photograph at a' with resulting camera co-ordinates x_a', y_a', z_a', which is further out from the centre of the photograph than the image at a. This change in position on the photograph due to the change in height of the object and the central perspective geometry is known as relief displacement. This causes tall objects such as office blocks and chimneys to appear to lean over when imaged on an aerial photograph. The effect also occurs in close-range photographs with changes in depth on the object, but may not be so readily apparent. Three projective transformation equations can be written for A' and a':

$$X_A - X_O = \lambda_a'. x_a'$$
$$Y_A - Y_O = \lambda_a'. y_a'$$

(2)

$$Z_A' - Z_O = \lambda_a' \cdot z_a$$

where

$$\lambda_a' = \frac{Z_A' - Z_O}{z_a} = \frac{(Z_A + \Delta Z) - Z_O}{z_a}$$

Comparing λ_a' with λ_a, it can be seen that the scale of the image at a' is larger than the scale of the image at a and it is this change of scale which causes relief displacement. Therefore, unless the object photographed is perfectly planar, it is not possible to scale accurate dimensions off a single photograph as the image scale across the photograph is constantly changing with the changing height of, or depth on, the object.

3.3.2 Tilted photography

In photogrammetry, it is extremely difficult to exactly align the camera axes with the object axes. Generally the camera is tilted with respect to the object co-ordinate system. Figure 3.3 illustrates the case where a photograph with a rotation of ω about the x axis produces a tilted camera co-ordinate system x_ω, y_ω, z_ω, so that an object point A is imaged at a. Also shown is the corresponding situation for an untilted photograph with the same perspective centre and camera axes system x_O, y_O, z_O and A imaged at a'. It can be seen that the y-plate co-ordinate on the tilted photograph, distance pa, is much greater than the corresponding co-ordinate on the

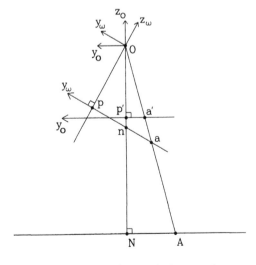

Figure 3.3 Relationship between tilted and vertical photography.

untilted photograph, $p'a'$. This change in position on the photograph is known as tilt displacement and is present on most photographs along with relief displacement.

To be able to use the projective transformation equations in (1) it is necessary to transform the tilted camera co-ordinates to their corresponding untilted co-ordinates. For the $(0$ rotation in Figure 3.3 the transformation would be :

$$x_O = x_\omega$$

$$y_O = y_\omega \cos \omega - z_\omega \sin \omega$$

$$\text{and } z_O = y_\omega \sin \omega + z_\omega \cos \omega$$

This can be written in matrix notation as:

$$\begin{bmatrix} x_O \\ y_O \\ z_O \end{bmatrix} = \begin{bmatrix} 1 & 0 & 0 \\ 0 & \cos\omega & -\sin\omega \\ 0 & \sin\omega & \cos\omega \end{bmatrix} \cdot \begin{bmatrix} x_\omega \\ y_\omega \\ z_\omega \end{bmatrix}$$

Now consider the secondary rotation 9 around the tilted y axis:

$$\begin{bmatrix} x_\omega \\ y_\omega \\ z_\omega \end{bmatrix} = \begin{bmatrix} \cos\varphi & 0 & \sin\varphi \\ 0 & 1 & 0 \\ -\sin\varphi & 0 & \cos\varphi \end{bmatrix} \cdot \begin{bmatrix} x_{\omega\varphi} \\ y_{\omega\varphi} \\ z_{\omega\varphi} \end{bmatrix}$$

Finally, consider the tertiary rotation K around the doubly-tilted z axis

$$\begin{bmatrix} x_{\omega\varphi} \\ y_{\omega\varphi} \\ z_{\omega\varphi} \end{bmatrix} = \begin{bmatrix} \cos\kappa & -\sin\kappa & 0 \\ \sin\kappa & \cos\kappa & 0 \\ 0 & 1 & 0 \end{bmatrix} \cdot \begin{bmatrix} x_{\omega\varphi\kappa} \\ y_{\omega\varphi\kappa} \\ z_{\omega\varphi\kappa} \end{bmatrix}$$

Normally the camera is tilted around all three axes so the transformation from the tilted to the corresponding vertical plate co-ordinates involves a rotation matrix resulting from the sequential rotations of ω, φ, and κ. This is usually written as:

$$\begin{bmatrix} x_O \\ y_O \\ z_O \end{bmatrix} = \begin{bmatrix} m_{11} & m_{21} & m_{31} \\ m_{12} & m_{22} & m_{32} \\ m_{13} & m_{23} & m_{33} \end{bmatrix} \cdot \begin{bmatrix} x_a \\ y_a \\ z_a \end{bmatrix} \qquad (3)$$

where x_a, y_a, z_a are the tilted plate co-ordinates; x_O, y_O, z_O are the corresponding vertical plate co-ordinates; and m_{11} to m_{33} are the nine elements of the three-dimensional orthogonal rotation matrix resulting from the sequential rotations ω, φ, and κ.

Combining equation (1) with equation (3) gives:

$$X_A - X_O = \lambda_a (m_{11}x_a + m_{21}y_a + m_{31}z_a)$$

$$X_A - Y_O = \lambda_a (m_{12}x_a + m_{22}y_a + m_{32}z_a)$$

$$Z_A - Z_O = \lambda_a (m_{13}x_a + m_{23}y_a + m_{33}z_a)$$

which, in matrix notation, is:

$$\begin{bmatrix} X_A - X_O \\ Y_A - Y_O \\ Z_A - Z_O \end{bmatrix} = \lambda_a \begin{bmatrix} m_{11} & m_{21} & m_{31} \\ m_{12} & m_{22} & m_{32} \\ m_{13} & m_{23} & m_{33} \end{bmatrix} \cdot \begin{bmatrix} x_a \\ y_a \\ z_a \end{bmatrix}$$

Rearranging gives:

$$\begin{bmatrix} x_a \\ y_a \\ z_a \end{bmatrix} = \frac{1}{\lambda_a} \begin{bmatrix} m_{11} & m_{21} & m_{31} \\ m_{12} & m_{22} & m_{32} \\ m_{13} & m_{23} & m_{33} \end{bmatrix} \cdot \begin{bmatrix} X_A - X_O \\ Y_A - Y_O \\ Z_A - Z_O \end{bmatrix}$$

Dividing the first two lines by the third line and substituting -f for Za gives:

$$x_a = -f \frac{m_{11} (X_A - X_O) + m_{12} (Y_A - Y_O) + m_{13} (Z_A - Z_O)}{m_{31} (X_A - X_O) + m_{32} (Y_A - Y_O) + m_{33} (Z_A - Z_O)} \tag{4}$$

and

$$y_a = -f \frac{m_{21} (X_A - X_O) + m_{22} (Y_A - Y_O) + m_{23} (Z_A - Z_O)}{m_{31} (X_A - X_O) + m_{32} (Y_A - Y_O) + m_{33} (Z_A - Z_O)}$$

The two equations in (4) are known as the collinearity equations and state that the object point, A, the perspective centre, O, and the image point, a, lie on a straight line, i.e. they are collinear.

3.4 Use of the collinearity equations

3.4.1 Space resection

Unless the camera is set up over a predetermined point and pointed in a known direction, a procedure which is very difficult to carry out precisely and is impossible in aerial photography, it is necessary for the position and orientation of the camera to be computed from the measurement of images of points whose object point co-ordinates, X Y Z, have been determined by some other method. These points are known as control points. This technique is called space resection and is similar in principle to the standard surveying resection technique carried out with a theodolite. However, the solution is three-dimensional, whereas in geodetic surveying a plane resection is normally only two-dimensional. For each image point two collinearity equations can be formed as observation equations, one for the x plate measurement, X_a, and one for the y plate measurement, y. There are six unknowns in a space resection – the co-ordinates of the perspective centre, X_O Y_O Z_O, and the three rotations, ω φ x, which make up the nine elements of the rotation matrix. The six unknowns are often referred to as the camera parameters and are the six degrees of freedom mentioned in 3.2.2. Therefore at least three control points with known object point co-ordinates, X_A Y_A Z_A, must be imaged on the photograph for a solution to be possible. As in all aspects of surveying it is usual to have redundant observations so four or more control points should be used and these control points should be well distributed about both the object area and the image area.

3.4.2 Reprojection

If the six camera parameters are known, from a space resection or otherwise, it is then possible to find the object co-ordinates of unknown points which are imaged on the photographs. From the collinearity equations there are now three unknowns – the object coordinates of the unknown point, X_A Y_A Z_A. However, the image point will yield only two observation equations, one each for Xa and Ya. Therefore *at least two* photographs upon which the unknown point is imaged are required to enable all three co-ordinates of the point to be found. Normally two photographs are used, referred to as a stereo pair, but more photographs will give a greater redundancy. Only one photograph is required if one of the three co-ordinates of the object point is known. This is possible in special cases when the object photographed is completely flat or the, say, Z co-ordinate of the point is known.

3.4.3 Phototriangulation

When the number of photographs required to cover an object increases beyond two it begins to become inefficient to have to supply three or more co-ordinated control points for each photograph for the space resection. In this case a technique known as phototriangulation may be employed.

Phototriangulation methods utilising the collinearity solution are called bundle adjustments.

For each photograph taken of the object, a bundle of rays can be recreated from measurements of the plate co-ordinates of the image points. A bundle adjustment links these individual ray bundles together and calculates the object co-ordinates of the imaged points and the camera parameters for each photograph. For each measured image point two observation equations can be derived from the collinearity equations. If the co-ordinates of the object point are known then the observation equations will be the same as those for a space resection. If the object point co-ordinates are not known then the two observation equations are modified to include the object co-ordinates among the unknowns along with the six camera parameters. Most bundle adjustment packages are flexible enough so that partially known object points, e.g. levelled points where Z is known but not X and Y, can be included and the observation equations arranged accordingly. As long as there are at least the same number of observations, i.e. plate measurements, as unknowns in the block then a solution is possible. As there are still six unknown camera parameters per photograph at least three image points must be measured on each photograph, but usually four or more well-distributed points are measured to add redundancy to the adjustment.

Object points whose co-ordinates are unknown must be measured on two or more photographs in the block. Also within the block there must beat least seven known object co-ordinates to provide a datum (X Y Z), scale (λ) and orientation around all three axes (Ω, Φ, K) for the block as a complete unit. More than seven known co-ordinates should be provided to give a suitable redundancy in the solution.

Therefore a bundle adjustment approach can greatly reduce the number of control points which have to be co-ordinated on site from at least three per photograph to a theoretical, but not recommended, minimum of three points for the whole object thus greatly reducing the time spent on site carrying out a conventional control survey.

The other major use of bundle adjustment procedures, in addition to greatly reducing the amount of on-site measurement for control, is in applications which require both high precision and high reliability. In the conventional reprojection from a stereo pair, the four observation equations from the two photographs for the three unknowns of the object point co-ordinates yield a redundancy of only one. It is a long-established principle in surveying that co-ordinating points by theodolite intersection produces significantly higher precision if the point is observed from three or more stations than if it were observed from only two. The same applies in photogrammetry. Also using three or more photographs not only increases the redundancy, but also enables the geometry of the spatial intersections to be improved. In many engineering applications highly convergent photography is used to improve the results compared with the standard stereo pair. When three or more photographs are used to co-ordinate each point the term multistation photogrammetry is often used to distinguish it from a stereo pair. However the increase in precision and reliability obtained using multistation photography has to be offset against the reduced number of points that can be measured on the object. When only two photographs are used they are usually viewed and measured stereoscopically. In this way, as long as the object has sufficient texture on it to enable the two images to be fused by the brain to form a three-dimensional image, then any number of points on the object can be measured. In multistation work, however, three or more photographs cannot be viewed and measured simultaneously and so the problem of identifying the exact same point on all the photographs, often taken from widely differing viewpoints, can degrade the accuracy of the solution. In this instance it is normal to affix specific targets to the object so that this problem of point identification is overcome. Obviously only a finite number of points can be targetted. This is not usually a major problem in monitoring applications as it is common practice to target the object being monitored anyway. Indeed, for high precision applications, targeting is mandatory.

3.4.3.1 Combined photogrammetric and geodetic bundle adjustments

In many uses of photogrammetry, and almost always in aerial mapping, the normal practice is to regard the conventionally surveyed control points as fixed and error-free and fit the photogrammetric measurements to this control. However in a number of high precision applications the inherent accuracy of the photogrammetric measurements are degraded if this approach is taken. Plate co-ordinates can be measured to a few microns or better which at a photograph scale of, say, 1:100 represents a few tenths of a millimetre on the object. It is extremely difficult, but not impossible, to obtain this level of accuracy using geodetic surveying techniques. Therefore, in a situation like this, it is desirable to use the photogrammetric measurements to improve the survey rather than use the results of the survey to degrade the photogrammetry. There are two ways in which this can be carried out. First, the survey observations can be adjusted in a normal survey adjustment package, preferably one that adjusts all three co-ordinates simultaneously, which will yield the adjusted co-ordinates of the control points and their variance matrix. This is then used as input into a bundle adjustment that allows the control points to be adjusted as well as all other points and camera stations using the variance matrix of the survey adjustment to constrain the movements of the control points. In this way a 'best fit' overall solution will be achieved with correct weighting being assigned to the survey results and the photogrammetric measurements.

The other solution to the problem is to enter all the survey and photogrammetric measurements together into one overall bundle adjustment program and adjust the combined networks as one homogeneous unit. The program must be arranged so that all the measurements have been assigned the appropriate weights.

It should be noted that both of these methods should give the same results for the same set of measurements.

3.5 Photographic media

3.5.1 Conventional film

If the high accuracy which close-range photogrammetry is capable of producing is to be exploited fully then as much attention must be paid to the photographic media being used as to any other aspect of the project. In traditional photogrammetry silver halide photographic emulsions are used. The emulsion has to be supported on a base material which is either a flexible plastic film or a plane glass plate. Glass plates offer better stability but the handling at large quantities of large plates can be a problem.

Dimensional stability of the base material is of prime importance. If the camera is fitted with a reseau plate and an analytical solution is to be employed, then significant film distortion can be allowed for by transforming the measured images of the reseau marks back to their calibrated values.

The unflatness of the base material is also very important and is much harder to model. To keep unflatness to a minimum the camera should have some form of film-flattening device. This can be in the form of either a vacuum back with pump to hold the film flat or a pressure plate system which presses the film forward on to a plane glass plate mounted in the focal plane of the camera. If glass plates are used then unflatness of the base material is of much less import importance. However even glass plates can bend under the strain of the film emulsion expanding or contracting with changing humidity. Therefore if the highest precision is required then extra thick plates should be used.

3.5.2 Digital imaging

Conventional photographs on film are often referred to nowadays as 'analogue' or 'hard copy' images. This is to differentiate them from 'digital' or 'soft copy' images. A digital image consists of a computer file containing a series of numbers (digits) which are interpreted by the computer to produce an image on the computer screen. Each number in the file describes the characteristic of a single pixel or picture element of the complete image. This number is a measure of intensity or brightness of that individual part of the image. For a black and white image this is the greyscale value and is normally a number between 0 and 255, where 0 refers to a pure black pixel, 255 a pure white pixel and numbers in between represent various shades of grey with increasing lightness with increasing intensity value. The complete picture is made up of a two-dimensional array of many hundreds, thousands or even millions of these pixel values. Figure 3.4 shows a typical digital image (a), an enlargement of an area of a around the target (b) and the corresponding array of pixel intensity values for b in (c). For a colour image, the file contains three sets of intensity values for each pixel, one for

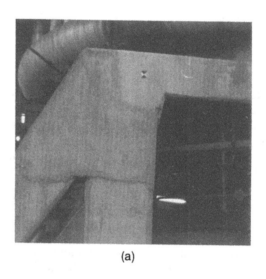

(a)

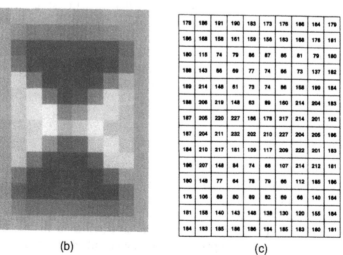

(b)

(c)

179	186	191	190	183	173	176	186	184	179
186	168	158	161	159	156	163	168	176	181
180	115	74	79	86	87	85	81	79	180
188	143	66	69	77	74	66	73	137	182
189	214	148	61	73	74	86	158	199	184
188	206	219	148	63	89	160	214	204	183
187	205	220	227	186	178	217	214	201	182
187	204	211	232	202	210	227	204	205	186
184	210	217	181	109	117	209	222	201	183
188	207	148	84	74	68	107	214	212	181
180	148	77	64	78	79	66	112	185	186
175	106	69	80	89	82	69	66	140	184
181	158	140	143	148	138	130	120	155	184
184	183	185	186	186	184	185	183	180	181

Figure 3.4 (a) Section of a typical digital image; (b) enlargement of target seen in (a); (c) array of pixel intensity values for image area in (b).

the red band, one for the green band and one for the blue band of the colour image. Hence a colour image containing the same number of pixels as a black and white image takes up three times the amount of disk space.

Figure 3.5 illustrates the problem of pixel size or image resolution. This shows four images of the same target with the resolution of each succeeding image being half that of the previous, i.e. the pixel size of each succeeding image is double the previous image. It can been seen that very soon the subject of the image becomes unrecognisable and any measurement taken from such an image would be very unreliable. This problem will be discussed further later.

One advantage of using digital images over conventional images is that image processing techniques can be employed. Most of these are beyond the scope of this book, but a number can be very useful. Image enhancement is a very powerful tool for improving the quality of a poor image. Sometimes it is not possible to obtain a good quality image because of, say, poor lighting conditions and the requirement for short exposure times because of vibration. With conventional photography this can result in an image which is very difficult or even impossible to measure because detail cannot be seen in the dark areas. However using a digital image it is possible to manipulate the pixel intensity values and increase the differences in intensity between adjacent pixels. This is known as contrast stretching. In a

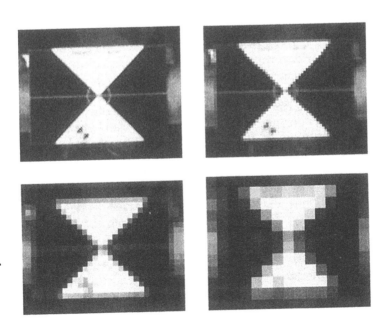

Figure 3.5 Images of typical survey target showing the effect of decreasing image resolution.

poor image such as in Figure 3.6a, the area of interest in the centre of the photograph may only contain pixels with intensity values in the range 10 to 60. This range of 50 intensity values can be stretched over the entire range of 0 to 255 so that a greatly improved image, Figure 3.6b, results. Other image processing techniques will be discussed later.

3.6 Cameras

As the photograph is the major unit of data in photogrammetry, it is of vital importance that the camera selected for a particular project is capable of producing an image of sufficient quality for the required task. Figure 3.1 and the collinearity equations refer to plate and camera co-ordinate systems and in the majority of cases in photogrammetry it is necessary for the camera employed to define these co-ordinate systems on the photograph. An exception to this rule is discussed later. Cameras which are capable of supplying this information are called metric cameras and cameras which do not define these co-ordinate systems are known as non-metric cameras. It is possible to modify non-metric cameras so that they can be used for measurement although they generally will produce results of a lower precision than is possible with metric cameras.

Figure 3.1 shows that the angle subtended at the perspective centre, O, by A and B is θ. In an ideal projection, the angle subtended at O by a and b, θ′, will also be θ. However, no lens system is absolutely perfect and the three points A, O and a, will not be exactly collinear as the ray will be slightly bent as it passes through the lens. This bending is known as lens distortion. If the distortion is small it may be neglected in all but high precision applications. If it is large, however, it must be allowed for in the solution or else the results will be virtually meaningless.

Before discussing cameras in more detail it is necessary to define a number of commonly used terms for which reference should be made to Figure 3. 1. The principal axis of the lens is defined by the ray which passes through the perspective centre, O, and intersects the film plane normal to it. This point of intersection, p, is called the principal point of the lens and is the origin of the plate co-ordinate system. The distance, Op, between the perspective centre and the principal point is called the principal distance or, more commonly, the focal length, f, of the lens and this is the z-co-ordinate of the camera coordinate system.

3.6.1 Metric cameras

In a metric camera all the information described above is determined by calibrating the camera after manufacture. Built into the camera must be some system which can be used to define the position of the principal point, p. This can be done by fixing in the focal plane of the camera a series of

Figure 3.6 (a) Typical low contrast image due to poor lighting conditions; (b) image in (a) after undergoing contrast stretching.

fiducial or index marks, normally four or eight. These are positioned either in the corners or midway along the sides of the format such that two lines drawn between the pairs of opposite fiducials will intersect at, or very close to, the principal point. The intersection of these two lines produces the fiducial centre. Any discrepancy between the fiducial centre and the principal point can be determined during the calibration and noted on the manufacturer's calibration certificate. Subsequent measuring of the plate co-ordinates with respect to the fiducial centre can be corrected to the principal point using this information. The other method of defining the principal point is to fit a glass plate in the focal plane of the camera with a series of crosses engraved on the glass which will be imaged on the photograph. This is a reseau and the central cross should be close to the principal point. Again any slight discrepancy is determined during calibration and allowed for in any subsequent measurements. In digital cameras, it is usually not necessary to have fiducial marks built into the camera. The normal practice is to use the comers of the CCD array.

The focal length, f, also needs to be known to a high degree of accuracy. In all air survey cameras the lens is of the fixed focus type and is set at a taking distance of infinity. This is acceptable as the aircraft flying height is so high that the ground being photographed can be regarded as almost flat and infinitely far away. In close-range photogrammetry the object photographed is obviously not infinitely far away and so cameras which are not focussed at infinity are required. These can be either of the fixed focal length type or of the variable focal length type. All metric cameras image the focal length on each exposure.

Metric cameras should have a minimal degree of lens distortion and what distortion is present in the lens should be calibrated so that even this small amount can be allowed for in high precision applications.

The necessity for maintaining the manufacturer's calibrated values for the principal point, focal length and lens distortion over a reasonable period of time requires construction of the camera to very high standards of stability and, in the case of variable focus cameras, repeatability. Any miscentring of the lens or change in the focal length over time will lead to errors in the measured results. Therefore metric cameras are built to the same high standards as all other pieces of surveying equipment and are, as a consequence, expensive. A typical metric camera for close-range applications is shown in Figure 3.7.

3.6.2 Non-metric cameras

There is always great pressure in photogrammetry to use non-metric cameras as they are significantly cheaper than their metric counterparts. However before succumbing to this pressure it is very important to know the limitations of these cameras and hence their limited applications. Non-

Figure 3.7 Zeiss UMK 10/13 IS 100 mm focal length film-based metric camera.

metric cameras will not have a fiducial reference system. The comers of the format can be used, but the 'fiducial centre' found from them may be significantly different from the principal point. The focal length is not known to a high precision and is variable and not repeatable even when the same focus setting is selected. A film-flattening device is not normally fitted. Additionally non-metric cameras have lenses designed for optimum picture quality and lens geometry and hence distortion is of secondary importance resulting in some lenses with very large distortions.

One way in which the low cost of these cameras can be utilised is to modify them to so-called 'semi-metric' cameras which are actually true metric cameras but of low precision. This normally involves fitting some fiducial reference system, usually a glass reseau, into the focal plane of the camera and permanently fixing the focal length of the lens. This is often done by gluing, taping or attaching a grub screw on the lens at a particular focal setting. This is necessary as the lenses used in non-metric cameras are not engineered to the precision of metric cameras and as the lens is focussed backwards and forwards the principal point of the lens can move across the image plane. Additionally when the lens is returned to a particular focal setting the focal length may not be identical to when it was set at that setting previously. The camera is then calibrated so that all the necessary information is known and the camera is then used as a normal metric camera. However, as the basic camera does not have the inherent stability of a purpose-built metric camera it should be calibrated at regular intervals.

If there are enough control points on the object then non-metric cameras can be used directly by employing a technique called self-calibration. In this the collinearity equations are expanded to include further unknowns for the displacement of the principal point, the focal length and lens distortion. In this system the camera is calibrated for every individual photograph used. However, this requires many more object control points than when using metric cameras and hence the cost saving of using a cheaper camera may be negated by the requirement for more fieldwork.

Cheap cameras can be used in photogrammetry, but they should be employed with care and should not be used at all if high precision is required.

3.6.3 Digital cameras

Recent years have seen the development of digital cameras. These are cameras in which the conventional photographic material has been replaced by an electronic light sensitive sensor. This sensor normally takes the form of an arrangement of charge coupled devices (CCDs). A CCD works by converting the photons which fall on it into electrons which are then accumulated in capacitors and converted into digital form which can be stored in and manipulated by a computer. For a digital camera to be of any use in photogrammetry it is important that the individual CCD elements are sufficiently small and in significant numbers to produce a high resolution image. The arrangement of the CCDs can be in either a single linear form or a two-dimensional array. If a linear array is used it is necessary for the array to sweep across the image plane of the camera to build up the final image. This requires both the camera and the object to be stationary for a significant period of time. However, a large two-dimensional array is able to capture an image in one short exposure. For example the Kodak DCS 460 uses a two-dimensional CCD array of 3,060 by 2,036 elements with a pixel size of 9 gm. This results in a black and white digital image of 6 Mb in size or a colour image of 18 Mb in size. The DCS460 is one of the highest resolution digital cameras currently available for 'professional' use, i.e. used for high quality for magazines, calendars, etc. However, because of its high resolution it is becoming widely used for photogrammetric purposes.

When a camera like the DCS460 is used to take a photograph the image is normally written to a PCMCIA card in a highly compressed format. The card is removed from the camera and downloaded to a PC. TWAIN compliant software on the PC decompresses the image which can then be manipulated or measured in numerous image processing packages.

Another series of digital cameras which has been extensively used in photogrammetry is the MegaPlus range from Kodak. These cameras differ from the DCS range mentioned above in that they are designed for laboratory use and are targeted for applications such as medical imaging, machine vision, metrology and microscopy. They have no on-board storage capacity and images have to be captured from the camera via an RS232 link with a PC fitted with a special frame-grabber board. As this range of cameras are restricted in their use, their application in structural monitoring has been confined to laboratory testing.

The above digital cameras, along with most others, are not designed as metric cameras. However, tests have shown that they maintain a remarkable stability although it is normal to permanently fix the lens at a particular focal setting using either a grub screw, strong tape or glue. It is also possible to increase stability further by rigidly fixing the CCD array in the back of the camera with a touch of glue.

There are now available a number of high resolution digital cameras, which have recently been developed specifically for photogrammetric purposes. Imetric of Porrentruy, Switzerland currently offer two models in their ICam Image Metrology Camera series, the ICam6 (Figure 3.8) and the ICam28. The ICam6 has an array of 3,072 by 2,048 pixels producing an image made up of just over 6 million pixels and the ICam28 has an array of 7,168 by 4,048 pixels resulting in a 29 million pixel image. They have specially designed lenses, an integrated flash unit, and an in-built Pentium-based PC. The on-board microprocessors are used to carry out various image processing tasks such as image compression thus reducing the quantity of data which needs to be transferred to other computers for further analysis.

Geodetic Services, Inc. of Melbourne, Florida, USA, a long-established company specialising in industrial photogrammetric systems, produce the somewhat similar INCA camera range. These utilise a Kodak MegaPlus camera which has been combined with an integral Pentium PC.

For a short time Zeiss produced a scanning back for their UMK range of conventional film cameras. This back replaces the normal film or plate back with a unit containing four CCD arrays, each of which is moved across a quarter of the total image format. The resulting four images are

Figure 3.8 Imetric ICam6 high resolution photogrammetric digital camera. (Photograph courtesy of Imetric SA).

joined together to produce a single image of 15,414 by 11,040 pixels of 11 µm size. Thus this camera can produce a very large high resolution digital image, but its main drawback is that it requires several minutes to complete the scanning process thus requiring a very stable camera platform and an object which does not move during the period of exposure.

One alternative to using a digital camera is to scan a conventional photographic image. There are numerous low-cost desktop scanners available on the market, few of which are suitable for photogrammetric purposes. This is because they suffer from a low resolution and/or a lack of geometric stability. To overcome this problem photogrammetric quality scanners must be used. These consist of either a linear CCD array or a two-dimensional array which is used to scan the required image section by section under high magnification. The sections are then joined together by software to produce a final image of high resolution. One such scanner is shown in Figure 3.9a and b. Typical photogrammetric quality scanners have an image resolution which can be varied between 5 and 100 µm and a geometric stability of 1 to 2 µm.

One major problem with digital images is their sheer size. As stated above a monochrome DCS 460 image requires 6 Mb of disk storage and a typical UMK photograph scanned at a resolution of 20 µm requires over 50 Mb of disk storage and a UMK photograph scanned at 10 µm requires over 200 Mb. It is possible to use image compression software to replace the disc storage required. However there is a danger that the image quality will be lost if too high a compression ratio is selected. Therefore selection of image compression ratio has to be balanced with the requirement to maintain image quality.

(a)

(b)

Figure 3.9 (a) LH Systems DSW500 high resolution photogrammetric film scanner; (b) interior of DSW500 showing scanning mechanism. (Photographs courtesy of LH Systems LLC).

3.7 Equipment for the measurement of photography

There are two forms of computer which can be used to carry out the task of solving the photogrammetric problem. One solution is to use mechanical analogue computers and the other is to solve the problem analytically, normally with a digital computer or calculator. In 1901, Pulfrich invented the stereocomparator which enabled plate co-ordinates to be measured. The resultant transformations from plate to object co-ordinates were carried out mathematically using tables and so were very tedious. To speed up the measuring process mechanical computers were developed and the first analogue stereoplotter was built in 1911. However with the development of the digital computer analytical techniques have been revived and now dominate.

The majority of instruments used today for the analysis of photographs are normally classed as either 'analytical' or 'digital'. This may appear confusing as both methods involve analytical techniques and digital computers! However this terminology is a result of historical developments in photogrammetry beginning with the trend away from analogue techniques in the early 1970s. It is hoped that the differentiation between analytical and digital systems will become clearer in the following sections.

3.7.1 Analytical methods of image measurement

The so-called analytical instruments involve the measurement of conventional photographic images, often the original negatives, on either film or glass. They are two dimensional measuring machines which are used to derive values for plate co-ordinates. The two main classifications of these instruments are comparators and analytical plotters.

3.7.1.1 Comparators

The simplest form of analytical instrument for photogrammetric measurement is the monocomparator which is basically an x-y measuring microscope. The photograph is placed on a stage plate and the microscope or, in some instruments, television camera is tracked across the photograph to the required image point. In other instruments the viewing head is fixed and the stage plate is moved. This produces machine co-ordinates which have to be transformed into plate co-ordinates and then to object co-ordinates. Because only one photograph is measured at a time only well-defined or targeted points can be measured. This is necessary so that the same point can be measured independently on two or more photographs.

The relationship between plate co-ordinates, x y, and machine co-ordinates, X_c Y_c, is shown in Figure 3.10 and can be expressed:

$$X_c = a_o + x \cos\alpha - y.\sin\alpha$$
$$Y_c = b_o + x.\sin + y.\cos\alpha$$

If the photograph has suffered from deformation (3.5.1) such as film shrinkage then an affine transformation may be used:

$$X_c = a_0 + \lambda_x.x.\cos\alpha - \lambda_y.y.\sin(\alpha + \beta)$$
$$Y = b_0 + \lambda_x.x.\sin\alpha + \lambda_y.y.\cos(\alpha + \beta)$$

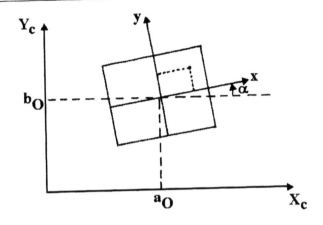

Figure 3.10 Relationship between plate and machine co-ordinates.

where λ_x and λ_y are the scale factors along the x and y plate axes, respectively, and β is the non-orthogonality between the x and y axes. Additional terms can be added to allow for the effects of the unflatness of the film or plate. Corrections for lens distortion can also be included.

The stereocomparator provides more flexibility by incorporating two stage plates and a binocular stereoviewing system. In this way untargeted points can be measured stereoscopically by simultaneous observation on two photographs.

3.7.1.2 Analytical plotters

All comparators are point measuring devices and, although capable of very precise measurement, are not suited for continual surface measurement. For this reason the analytical plotter was developed to combine the high precision of a stereocomparator with the continuous measurement facility of an analogue stereoplotter. The invention of the analytical plotter is credited to U.V. Helava in 1957, but it was not until 1976 that the analytical plotter really made an impact in commercial photogrammetry.

In general appearance the analytical plotter (Figure 3.11) looks like a stereocomparator and consists of a stereo binocular viewing system, two stage plates and a controlling computer. The main difference between the two types of instruments is the way in which the photographs are moved in the machine.

In a comparator, there is a direct mechanical connection between the operator input (normally via handwheels) and the movement of the photographs. This normally takes the form of a worm and gear drive system. In an analytical plotter this mechanical connection has been removed and replaced by an electronic connection. The operator's inputs are sensed,

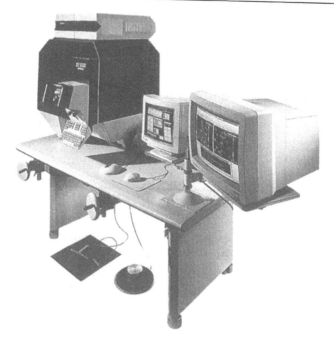

Figure 3.11 LH Systems SD3000 analytical stereo plotting instrument. (Photograph courtesy of LH Systems LLC).

either by rotary encoders attached to handwheels or a digitising tablet below a sensitised mouse. These inputs are transmitted to the digital computer which computes how the photographs should be moved and then sends commands to servomotors attached to the stage plates. This procedure is known as the 'real-time loop' and is normally carried out around 50 times per second so that, to the operator, it is a continuous smooth movement of the photographs.

Analytical plotters can be operated in either 'comparator mode' or 'plotter mode'. Comparator mode is used for the initial measurement of the fiducial marks to solve for the inner orientation and also for the measurement of the images of the control points for the solution of the space resection. During these processes, the inputs sensed from the operator are translated into similar movements of the stage plates.

Once the camera parameters have been computed the method of operation changes as the system switches into 'plotter mode'. Looking through the binocular viewing system the operator now sees a three-dimensional stereoscopic image of the object photographed.

Within the stereo image is a measuring mark which appears to move in three dimensions around the model, the so-called 'floating mark'. The operator controls the movement of the mark using either handwheels or a 3-

D mouse. In most models of analytical plotter the operator's inputs are sensed as if the operator was moving the floating mark around in the real world, i.e. the operator's inputs are interpreted as movements in the X, Y and Z object co-ordinate system. Using these co-ordinates and the solutions for the space resection and inner orientation the computer calculates what plate, and hence machine, co-ordinates correspond to these object co-ordinates and then drives the stage plates to these positions. Therefore these machines appear to work backwards, i.e. instead of measuring plate coordi-nates and computing object co-ordinates they actually measure object co-ordinates and compute plate co-ordinates! However, as the operator is viewing how the measuring marks move across the photographs the result is the same. If the fused floating mark appears to be off the surface or target being measured the operator adjusts the mark by changing its position in the object co-ordinate system. The computer moves the stage plates and the mark now appears to be at the desired position. An experienced photog-rammetric operator can very rapidly measure single point positions and can also plot continuous linear features such as roads, rivers and contours. This ability to specify object co-ordinates and drive to the corresponding plate co-ordinates can be a useful tool in some applications in structural monitoring. An example of the use of this feature is described in Chapter 4.

3.7.2 Digital methods of image measurement

In this case the term 'digital' is used to describe instruments or processes which use digital images as opposed to conventional film images. The pho-togrammetric computations are generally the same as those used in analyti-cal instruments.

In its simplest form, all that is required is a computer display screen and some method of extracting image co-ordinates off the screen as a mouse is used to move a cursor across the image. Many commercial image pro-cessing packages will have a facility for displaying the pixel co-ordinates of the position of the cursor. Typical packages (among many) are Microsoft Paint (bundled as standard with Windows), Paint Shop Pro and Adobe Photoshop. In Paint, the pixel co-ordinates are continuously displayed towards the bottom right of the window. In Photoshop they are displayed under 'Info' in 'Navigator'. Therefore simple packages like these (with Paint present on virtually every PC) can be used for the measurement of a digital image. Unfortunately when using these packages the pixel co-ordinates have to be written down by hand, a very time-consuming and tedious task. There is no facility to automatically record the co-ordinates. Also they do not contain any algorithms for computing object co-ordinates from the measurements. Analysis software would have to be purchased separately. However, they do illustrate very simply the concept of measuring co-ordinates on an image.

Until recently most digital photogrammetric systems were based on Unix workstations. However these have generally been superseded with systems based on high performance PCs. There are numerous general purpose systems available from major suppliers such as LH Systems and Z/I Imaging. The major market for these systems is aerial mapping. In order to meet this requirement they include stereo-viewing systems utilising either polarising or liquid crystal viewing spectacles. These systems involve rapidly alternating the display of two photographs on the screen. In the polarised viewing system this alternating is synchronised with a polarising filter mounted in front of the screen which switches between horizontal and vertical polarisation. The screen is viewed through polarised spectacles with one lens horizontally polarised and one lens vertically polarised. In the liquid crystal system the lens of the viewing spectacles are made of liquid crystal, which are alternately cleared and blanked in synchronisation with the alternating display on the screen. The result in both systems is that the viewer's left eye only sees the displayed left photograph of the stereo pair and the right eye only sees the right-hand photograph. This results in an apparent three-dimensional image which appears to stand out from the computer screen. This stereo image is measured using a 3-D cursor and 3-D mouse as in an analytical plotter.

This stereo viewing system is used in 'plotter' mode. In initial operation, as with an analytical plotter, the instrument operates in 'comparator' mode, where photographs are measured individually to recover the inner orientation and camera parameters. Once this has been done, 'plotter' mode is engaged and operation is identical to that described above for an analytical plotter, including the ability to move the photographic images to previously measured co-ordinates.

One major advantage of a digital system over an analytical system is that more than one person can view the images at the same time. In this way one or more engineers, who may not have sufficient experience of photogrammetry to be able to undertake accurate measurements themselves, can point out important features to an experienced operator who then takes the measurements. Measurement of single images is simple but it does require experience before reliable 3-D measurements can be taken from a stereo pair.

3.8 Automation of the photogrammetric process

Manual measurement of images can be time-consuming and, hence, expensive (although still cheaper than alternative non-photogrammetric solutions). Therefore much work has been done in attempting to speed up the measurement process using automation.

3.8.1 Digital image correlation

Digital image correlation is a technique employed to automatically or semi-automatically produce measurement from stereo pairs of imagery. In the semi-automatic method the operator approximately selects the same feature in both images and then instructs the computer to carry out the final exact measurement automatically. This is especially useful if the operator is relatively unskilled in stereo measurement. In the fully automatic process many hundreds or thousands of points in a pair of photographs are automatically measured by the system.

Digital image correlation involves a process of area-based matching in which a small area of the image is selected on one photograph and the computer attempts to find the same pattern of pixel intensities on the second photograph. Figure 3.12 illustrates this. A small area of 3 by 3 pixels is selected on one image, either automatically or by an operator. A larger area, say 7 by 7 pixels, is searched to find a similar pattern. If no matching pattern is found the search area in the second photograph is increased in size. This system works well if the area being measured is flat or gently undulating and has a lot of fine detail. If the area is featureless, i.e. all pixels have roughly the same intensity levels, then incorrect matching can occur. If the area is viewed from widely different viewpoints, as in many close-range applications, matching will also fail. This is because objects in the foreground will obscure different areas of the background resulting in two images which appear significantly different. Therefore manual measurement will still remain necessary for some applications for many years to come.

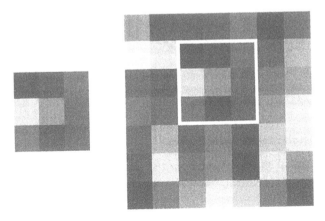

Figure 3.12 Principle of digital image correlation. 7x7 patch on right searched for match of 3x3 patch on left.

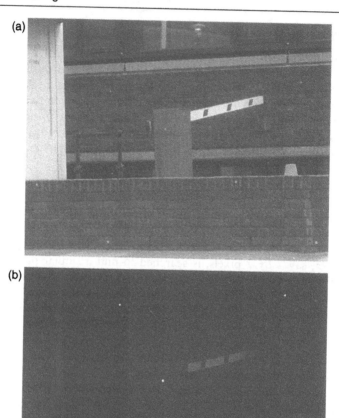

Figure 3.13 (a) Photograph taken under normal exposure and lighting conditions; (b) same area as in (a) taken with fast exposure and synchronised flash to show the use of retrotargetting.

3.8.2 Automatic target measurement

Although digital image correlation may have some applications in structural monitoring a much more important technique is automatic target measurement. This works in a similar fashion to digital image correlation except that instead of two images being compared, a single image is compared with a specified pixel pattern corresponding to the particular targets being used. For instance, if an object is targeted with a number of targets similar to the one shown in Figure 3.4b, each image is searched for the pixel pattern shown in Figure 3.4c and the pixel co-ordinate of the centre of the patch is taken as the image co-ordinate of that particular target.

This process can be greatly simplified and made more accurate if retro-targeting is adopted. Retrotargets reflect light back towards the source of origin of the light, similar in principle to a retroprism used with electro-magnetic distance measuring instruments. Retrotargets are often made from small circles of reflective tape. The camera has a flash mounted very close or, preferably, around the lens and a short exposure time is used. The resulting image may not show any detail of the object at all except for the targets. Figure 3.13a shows a photograph taken under normal exposure conditions where the circular targets can been seen as light grey dots. Figure 3.13b shows the same area taken with a very fast exposure with a synchronised flash. The circular retrotargets now stand out as bright dots against the very dark background. Various techniques such as thresholding and image masking can be used to determine the centre of each target on the image. As the targets are normally circular the image masking consists of mathematically fitting an ellipse to the image. This results in the determi-nation of the centre of the target to subpixel accuracy, often to one tenth of the pixel size. Therefore with a typical pixel size of 9 gm, the centre of the target can be determined to a precision of ± 0.9 μm. The image can be automatically searched and all target images identified and measured automatically. Therefore extremely precise image measurements can be achieved very rapidly, virtually in real-time, if the camera is directly attached to a PC. This has resulted in numerous industrial applications including real-time tracking of robotic manufacturing machines. Examples of applications in structural monitoring are given in the next chapter.

More detailed information on digital photogrammetry can be found in Atkinson (1996).

Reference

Atkinson, K.B. (ed.) (1996) *Close Range Photogrammetry and Machine Vision.* Whittles: Caithness.

Chapter 4

Photogrammetry – practice and applications

Donald Stirling

Photogrammetry is widely used in many fields of engineering, e.g. aerospace, automotive, shipbuilding and process, as well as for structural monitoring. The following examples will illustrate a number of projects where photogrammetry has been used for monitoring purposes. These examples will highlight the many advantages that photogrammetry can provide. One of the main advantages is that it is a non-contact measurement system, i.e. measurements can be undertaken of objects with difficult or restricted access, objects which are dangerous, and objects which are in motion. For this last case it is necessary to take all the necessary photography simultaneously in order that the motion can be frozen at one moment in time.

Another significant advantage of photogrammetry is that it is a very rapid method of data capture. Therefore large numbers of points on a structure may be monitored with minimal time on site. Additionally, conventional monitoring techniques can normally provide information on movement of a limited number of discrete points on the structure. Photogrammetry, on the other hand, records the entire structure enabling much more information to be extracted. However, if results of the highest precision are required it is still necessary to physically attach targets to the structure before the commencement of the work.

Photogrammetry also provides an archive record. This means that the measurement of the images need not be carried out immediately. For instance, if movement is suspected, a series of photographs can be obtained over a period of time. These photographs can be analysed with only a small number of points on the object being measured. If movement is detected then a more detailed analysis can be carried out. Conversely if movement is not detected then only minimal expense in measuring time will have been incurred. If conventional surveying techniques were employed all points on the structure would have to be measured at each epoch before the results could be analysed to ascertain whether the structure was moving or not. Therefore using photogrammetry can save a great deal of unnecessary time and expense.

These advantages will be illustrated in the following applications. These examples are of real monitoring surveys and are used to illustrate the broad range of structures which can be monitored. A description of the site techniques which were used is included which will highlight the flexibility of photogrammetry in a variety of situations.

4.1 Monitoring of a building facade

During the tunnelling works connected with the extension to the Jubilee Line underground railway in London a great deal of monitoring of buildings above the route of the tunnelling was undertaken. Much of this consisted of precise levelling to bolts drilled and cemented into various building in order to detect any surface settlement during the tunnelling process. However some surveys also included the use of photogrammetry to monitor entire facades. One such survey concerned a building adjacent to Waterloo Station. Three epochs of monitoring were carried out – each epoch timed to coincide with various stages of the driving of the two running tunnels beneath the building. The building concerned was a typical 1960s era office block composed of two adjoining sections, one of seven floors and one of ten floors. Before the first survey, a total of 56 high contrast targets were affixed to the face of building.

At each survey epoch all 56 targets were co-ordinated by geodetic survey methods, namely angular intersections measured with electronic tacheometres from five control stations positioned at street level opposite the facade. At each epoch, all geodetic survey observations were completed in a single day.

At the same time as the geodetic surveying was being carried out, a series of twenty-seven photographs were taken using a Zeiss UMK 10/1318 large format metric camera. Sample photographs are shown in Figures 4.1 and 4.2. The images were recorded on glass plates for maximum stability and to reduce the effects of unflatness in the camera image plane. This photography was taken to greatly improve the precision and reliability of the derived co-ordinates of the facade targets. When each photograph was measured and a bundle of rays produced mathematically it produced the equivalent of an additional geodetic survey station. The time required to observe with an electronic tacheometer from an additional twenty-seven geodetic stations would have been prohibitive, whereas the acquisition of the twenty-seven photographs was undertaken in less than 3 hours. The photographs were taken from ground level, from a footbridge and from the roofs of a number of buildings opposite the building being monitored. The roof photography in particular greatly improved the geometry of the combined geodetic and photogrammetric network.

After each epoch, the data from the memories of the electronic tacheometers were downloaded and an initial adjustment of the network carried out using a three-dimensional least squares adjustment program. The

Figure 4.1 Sample UMK plate of building close to Waterloo Station, London. Plate taken for monitoring purposes related to Jubilee Line Extension tunnelling works.

Figure 4.2 Sample UMK plate of building close to Waterloo Station, London. Plate taken for monitoring purposes related to Jubilee Line Extension tunnelling works.

photographs were measured on an analytical stereoplotter with a measuring precision of ±3 μm operating in comparator mode, i.e. the output was a file of plate co-ordinates for each photograph.

The least squares adjustment program was run again using the output target coordinates from the geodetic adjustment as the starting values for the new adjustment. The observations used for this new adjustment consisted of both the geodetic survey observations and the photogrammetric observations combined, with each observation being suitably weighted in the adjustment.

After the adjustments were carried out for all three epochs, an analysis was carried out to determine possible deformations in the facade of the building. The resulting deformations were subjected to a full error analysis. The result of these surveys appeared to indicate some differential settlement of the building. This differential settlement coincided with an expansion joint in the structure. A geodetic survey consisting of the five ground stations alone would not have yielded a result of sufficient precision and reliability to have confirmed these movements.

4.2 Monitoring of a stone railway viaduct

British Rail were concerned that, with the introduction into service of high-speed trains, a viaduct on the east coast main line north of York was deforming. Initial monitoring surveys consisted of precise levelling to metal brackets attached to the stonework to measure settlement and measuring offsets from a laser line shone between plinths at each end of the bridge to detect horizontal deflections. Unfortunately these methods required personnel working on the bridge itself which necessitated a line possession. Hence the line was closed while the measurements were being taken and trains diverted. As this is one of the busiest stretches of railway in the U.K., major disruption resulted. In order to reduce this disruption British Rail decided to try photogrammetry as a way of monitoring the bridge instead.

Targets were placed around the tops of the arches and along the level of the trackbed. A series of twelve photographs (Figure 4.3) were taken from an adjacent bridge which was used by freight trains. This required a line possession of this bridge. However if the possession was taken at a weekend then disruption was minimal as there were virtually no freight trains running at weekends at that time.

A geodetic survey network was observed around the bridge from a stable datum point. This datum point was some distance from the bridge as the entire river floodplain was susceptible to movement. Approximately 10% of the targets on the bridge were co-ordinated by theodolite intersections to provide the control for the photography. The photography was measured on a stereocomparator and a bundle adjustment carried out.

The result of three epochs of measurement proved that photogrammetry

Figure 4.3 UMK being used to photograph railway viaduct. Visible on the viaduct are the targets used for the monitoring. Also visible are metal brackets which had been used for precise levelling surveys previously.

was a suitable tool for three-dimensional deformation monitoring. Its principal advantage over the methods used previously was that it was non-contact and caused no disruption to rail traffic on the main line. Throughout the survey and photography field time trains continued to cross the bridge at the maximum line speed of 125 miles per hour (200 kph).

4.3 Monitoring of reservoir embankments

Figure 4.4 shows the interior lining of a reservoir where the inner embankment has slumped because of water penetration over a length of about 50 m. Movement of the embankment had been monitored for some time by conventional geodetic surveying methods. This had involved taking measurements from a datum pillar close to, but outside, the region of movement to a number of pillars constructed at approximately 10 m intervals in the area under study. However, this only provided information on movement at a very limited number of points and study of the photograph shows that the movement was more complex than was detectable using measurements to the pillars alone. Therefore it was decided to employ photogrammetry to provide more detailed information on what was happening. Photography was taken using a Zeiss UMK camera with a 100 mm focal length. Because of the width of the reservoir it was necessary to take the photography from a small dinghy looking up the embankment slope (Figure

Figure 4.4 Interior lining of reservoir which has suffered from water penetration and hence slippage.

4.5). This illustrates another advantage of photogrammetry over conventional surveying methods, namely the lack of requirement for a stable instrument station. It is only necessary for the camera to be free of movement during the very short time required to expose each photograph. Additional photography was taken from the rear of a Land Rover looking down the embankment slope (Figure 4.6). The Land Rover was used in order to raise the camera so that a better view could be obtained downslope.

Control was provided by using small black and white targets which could be rotated between the photography from the dinghy and the photography from the Land Rover (one such target is visible in Figure 4.6). These targets were co-ordinated by conventional geodetic surveying methods based on the existing datum pillar.

During the survey for the second epoch a 300 mm focal length UMK was available. The increased focal length enabled photography of a suitable scale to be taken from the far side of the reservoir thus eliminating the need for the dinghy. In addition the higher viewpoint obtained from the top of the opposite bank meant that photography looking down the slope from the Land Rover was also not required.

The photographs were measured on an analytical plotter. During the measurement of the photography for the first epoch, the corners of a large number of stone blocks where selected and co-ordinated. In this particular case the instrument was used in plotter mode as opposed to comparator mode, thus recording the real world 3D coordinates and not plate co-

Figure 4.5 UMK camera in small dinghy to photograph reservoir lining.

ordinates. The same block corners were then measured on the second epoch's photography and the movements determined. Figure 4.7 shows a plot of these measured movements. The movement in x and y is shown by a vector and the movement in z, i.e. settlement, is indicated by the radius of the circle at the end of the vector. When analysis of these movements was combined with borehole data, a better understanding of the mechanism of the slippage was obtained than was possible using the previous limited survey data.

4.4 Monitoring of an earth embankment dam

Figure 4.8 shows the downstream slope of an earth embankment dam which has a number of discrete areas of settlement in it. As in the reservoir lining example, the slope had been monitored by taking conventional geodetic survey measurements from datum pillars to targets mounted on a number of small concrete pillars dug into the face of the slope. Two of these pillars, and their wide spacing, are shown arrowed in Figure 4.9. As in the previous example this relatively small number of monitoring points was insufficient to detect isolated areas of movement between the pillars. Once again photogrammetry, using photographs taken with a 100 mm UMK (the camera is visible in Figure 4.9), was used to record the entire face of the embankment.

Figure 4.6 UMK mounted in Land Rover to photograph reservoir lining from above.

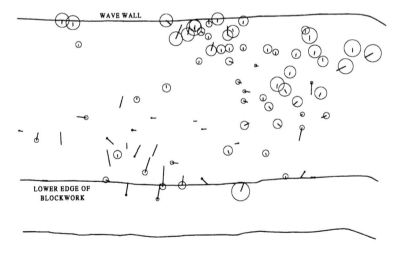

Figure 4.7 Plot of movements of reservoir lining.

Figure 4.8 Downstream face of an earth embankment dam which had suffered from isolated areas of settlement.

Figure 4.9 Section of earth embankment dam with monitoring pillars arrowed. Also visible is a Zeiss UMK camera being used to take photography of the dam.

However, the method used for measuring the photography was slightly different. Once again the photographs were measured as stereo pairs in an analytical plotter. This time the real world object co-ordinates for a regular grid of points across the embankment were recorded from the epoch 1 photography. The technique used for measuring the epoch 2 photography utilised a very useful feature of an analytical plotter. As described in 3.7.1.2 when the system is in 'plotter mode' the computer uses 3-D object co-ordinates to calculate the positions of the stage plates and drives them to that position. Normally the movements in the object co-ordinate system are provided by the operator turning handwheels or moving a mouse. However, it is also possible to use co-ordinates stored in a disk file to drive the stage plates. In this case the measured object co-ordinates of the surface points measured in epoch 1 were used to drive the machine. The operator monitored the movement of the floating mark in the stereomodel. If the floating mark rested on the surface of the dam then no detectable settlement had occurred at that point. However, if the mark appeared to lie off the surface then movement had occurred in the embankment. In this way all the measured grid points from epoch 1 were very rapidly measured on the epoch 2 photography and the large areas where no settlement had taken place noted. Where areas of settlement had been identified then a denser grid of points was measured on both sets of photography. Therefore the need to take a large number of measurements in areas were no movement had occurred was avoided.

4.5 Monitoring of a retaining wall

The A55 is a major trunk route in north Wales to the port of Holyhead on the island of Anglesey. Gradually over the last 20 years this road has been upgraded to dual carriageway. One of the last sections to be upgraded was a short stretch between Penmaenmawr and Llanfairfechan. The only feasible way to provide a second carriageway was to drive an 880 m long tunnel through the granite headland of Pen-Y-Clip. At each tunnel portal large retaining walls were constructed with the wall of the west portal being 220 m long and 35 m high, thus making it one of the largest retaining walls in Europe. The walls were constructed from the top down using 3 m by 2 m precast concrete panels with each panel being anchored to the rock by two tensioned rock anchors. It was required to monitor the behaviour of these walls both during and after construction. As the walls were being constructed on a very steep hillside, conventional geodetic surveying methods would not have been able to provide results of sufficient reliability because of the poor geometry of the network. Also as monitoring would be carried out during the construction phase it would have been almost impossible to have been able to take measurements to all targets with much of the wall being obscured by construction plant. The use of photogrammetry enabled superior network geometry to be obtained by raising the camera above the level of the existing road. One possibility would have been to employ an elevating platform. However, this would have resulted in severe congestion on the existing single carriageway road. Therefore it was decided to employ a helicopter to obtain oblique photography of the walls.

A Zeiss UMK camera with 300 mm lens and glass plates was used for the photography. It was also necessary to design a special mounting for the camera. This mounting had to be strong enough to support the weight of the camera, but also enable pointing and plate changing to be accomplished easily and quickly. The final design is shown in Figure 4.10.

Targets were rigidly fixed to selected panels on the walls with five targets on some panels, one in each corner and one in the centre, so that rotations of individual panels could be detected. Additional targets were fixed to the rock above the two portals, so that any movement in the rock could also be detected.

At each epoch three runs of photography were obtained from the helicopter. One run was positioned low down level with the base of the wall and at a distance of 90 m from the wall. The second run was positioned 100 m higher at a distance of 70 m from the wall. A third run was executed

Figure 4.10 300 mm focal length Zeiss UMK camera in specially constructed hand-held mount for photography from a helicopter.

at a mid-level position at a range of 150 m from the wall to provide an overall coverage. Each wall was photographed over a 10-min period. A 10-min limitation had been imposed by the local police as they had insisted that all traffic be halted when the helicopter was in the vicinity of the site. Examples of photography from each of these runs are shown in Figures 4.11, 4.12 and 4.13.

The photographs were measured on an analytical plotter operating in comparator mode with each target being measured three times. The resulting mean standard errors of the plate co-ordinates was the order of ±3 μ. The observations were processed in a least squares bundle adjustment program. As no ground survey had been carried out to co-ordinate control targets there was no fixed datum for the survey. The method of 'inner constraints' was adopted to overcome this lack of datum, i.e. the starting values for the co-ordinates of the targets were the same for each epoch's adjustment. The resulting deformations were determined to a precision of ±1.5 mm. A typical plot of movements with 95% confidence ellipses is shown in Figure 4.14.

Once the tunnel had been completed and the second carriageway opened, it then became possible to close the old carriageway to allow photography to be taken from a hoist instead of from a helicopter (Figure 4.15).

The above information and associated Figures are published with the kind permission of the Highways Agency of the U.K. Government.

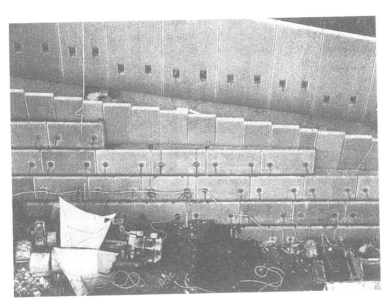

Figure 4.11 Typical UMK plate of A55 retaining wall from low-level run.

Figure 4.12 Typical UMK plate of A55 retaining wall from high-level run.

Figure 4.13 Typical UMK plate of A55 retaining wall from mid-level run.

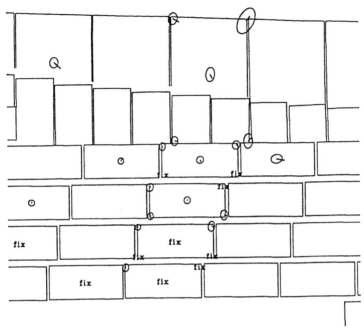

Figure 4.14 Plot of movements of A55 retaining wall.

Figure 4.15 Hoist being used for A55 retaining wall photography.

4.6 Monitoring of tunnels

The normal practice in photogrammetry is to have a minimum of two photographs of an area on an object so that the 3-D co-ordinates of points on the object can be photographed. However, it is sometimes possible to solve a particular problem using only a single photograph. One such case is in measuring tunnel profiles. A laser projecting through a rotating right-angled prism has been used to draw a vertical profile line around the tunnel wall. Using a long exposure, the image shown in Figure 4.16 has been generated. As the resulting laser line defines a flat plane then one co-ordinate value can be fixed and there are now only two unknown co-ordinates to be solved for using the collinearity equations. Hence a single photograph will allow a solution to be obtained. The photograph can be scaled in a number of ways ranging from a single known distance appearing on the photograph up to, in the example shown, full 3-D co-ordinates provided for a number of points on the profile. If the position of the profile is marked the laser can be repositioned at a later date and a second photograph obtained. Comparison between the two sets of measurements will reveal any deformation in the tunnel profile. This technique is also com-

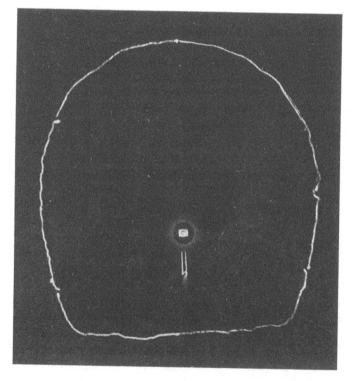

Figure 4.16 UMK plate of tunnel profile generated by rotating laser system.

monly used for determining clearance models of tunnels for new rolling stock. There are now real-time systems available where a CCD camera is used to track the laser to produce the profile instantly.

4.7 Laboratory testing

Photogrammetry has been widely used to measure the results of load testing in laboratories. Object sizes can vary from small samples of materials or models up to full size structures such as the UK's Large Building Testing Facility (LBTF) at Cardington. The number of photographs required per epoch can vary from one to tens of photographs. As in the case of tunnel profiling, it is possible to use a single photograph if movement in only one plane is expected. For instance if a beam is being loaded from above it is often reasonable to expect that all deformation in the beam will occur in the vertical plane. However, the use of at least two photographs would uncover any unexpected horizontal deflection of the beam. If complex 3-D deformation is expected then highly convergent multistation photography should be used. Figure 4.17 shows a sample of steel cable from a large suspension bridge which was tested in a fatigue rig. Targets were fixed on the rig framework to provide control for a single photo resection. The subsequent reprojection was on to a depth value set at the z-value of the targets on the front of the cable (movement of the cable was in the x,y plane). A similar test was carried out using close-up photography of one section of the cable where targets were attached to individual strands of

Figure 4.17 UMK plate of test rig for fatigue testing of suspension bridge cable.

the cable. This was used to detect any differential movement of the cable strands.

4.8 Applications of digital photogrammetry

All of the examples described above utilised film-based cameras, although measurement was sometimes carried out using digital equipment after the film had been scanned with a high resolution photogrammetric scanner. Using retro-targeting, the image measurement was automated in some cases. At the time of writing most applications of 100% digital photogrammetry for structural monitoring have been restricted to laboratory testing. This is because of the somewhat low image resolution of current digital cameras when compared with large film-based cameras such as the Zeiss UMK. However with continuing improvements in image resolution this situation could change in the near future. If retro-targeting is used combined with digital cameras attached directly to a PC, results can be obtained in real-time. Some examples of real-time laboratory applications are given below.

A system using four CCD cameras permanently set-up in an environmental chamber was used to monitor deformations of a large number of wood panels with changing temperature and humidity (Robson et al., 1995). The panels were representative of the materials used in mediaeval paintings which were gradually cracking and deteriorating because of unstable storage/display conditions. Each panel had over a hundred small retro-targets attached and were constantly monitored as the temperature and humidity were altered. The results from the system were used to ascertain the optimum atmospheric parameters to reduce further deterioration to a minimum.

Similar systems have also been used to rapidly capture deformations during the test loading of structural members such as concrete columns (Woodhouse and Robson, 1998) and reinforced concrete beams (Woodhouse et al., 1999). In both these examples multiple CCD cameras and retro-targeting were used not only to detect the deformation in the samples, but also to monitor the stability of the test rig to determine any influence this may have on the results of the test.

Because digital photogrammetry can produce deformation information in real-time it can also be used to actually control the experiment. This has already been used to monitor the movement of material samples in a geotechnical centrifuge. 3-D movements of the sample were detected in real-time and a feedback loop used to vary the speed of the centrifuge and hence the loading on the sample (Robson et al., 1998).

As digital camera technology improves and costs reduce, it is expected that it will be possible very soon to undertake real-time monitoring of full-size structures. Sets of images taken at regular intervals, varying from fractions of a second to hours or days apart, will be automatically measured and targets tracked between frames. There are currently a number of research teams working on developing useable systems which are expected to be available in the near future.

References

Robson, S., Brewer, A., Cooper, M.A.R., Clarke, T.A., Chen, J., Setan, H.B. and Short, T.M. (1995) 'Seeing the wood for the trees – an example of optimised digital photogrammetric deformation detection', ISPRS Intercommission Workshop: From Pixels to Sequences, Zurich, 22–24 March 1995. *International Archives of Photogrammetry* 30(5WI): 379–84.

Robson, S., Cooper, M.A.R. and Taylor, R.N. (1998) 'A digital imaging system for determining 3D surface displacements in geotechnical centrifuge models', Proceedings of the 11th International Conference on Experimental Mechanics, Oxford, UK, 24–28 August 1998, in I.M. Allisson (ed.). *Experimental Mechanics – Advances in Design, Testing and Analysis*. Rotterdam: A.A. Balkema, 1998, pp. 647–52.

Woodhouse, N.G. and Robson, S. (1998) 'Monitoring concrete columns using digital photogrammetric techniques', Proceedings of the 111th International Conference on Experimental Mechanics, Oxford, UK, 24–28 August 1998, in I.M. Allisson (ed.). *Experimental Mechanics – Advances in Design, Testing and Analysis*. Rotterdam: A.A. Balkema, 1998, pp. 641–6.

Woodhouse, N.G., Robson, S. and Eyre, R.J. (1999) 'Vision metrology and three-dimensional visualization in structural testing and monitoring', *Photogrammetric Record*, 16(94): 625–41.

Chapter 5

Surveying type monitoring and its place within a comprehensive structural monitoring system

David Cook

5.1 Introduction

This chapter deals principally with surveying in connection with tunnelling or excavation-related movements and associated mitigation measures. Many techniques will be applicable regardless of the mechanism causing the structural movement. Each type of structural monitoring has its own particular advantages and limitations and this chapter identifies the place of surveying within an integrated monitoring system. The focus is on approach rather than on particular instruments which would quickly become out of date (Figure 5.1).

Monitoring operations are considered to provide high accuracy data at specific, well-defined locations rather than more global techniques, such as photogrammetry. It is assumed that structural monitoring is looking to accurately detect movements ranging between 0 and 25 mm, with a resolution of 1 mm. The accurate measurement of submillimetre movements on

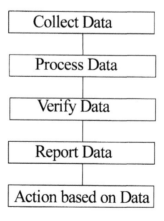

Figure 5.1 The monitoring process.

site is possible but expected derived benefits need to be carefully evaluated against the additional costs which will be incurred. For this precision the best survey techniques need to be used. This is to ensure that the systematic and random errors propagated in observations are less than the smallest displacement that is being measured or expected.

Regardless of the size of the monitoring works a clear understanding of what is required and how it is to be achieved needs to be in place prior to installation of a monitoring system. Generally monitoring data are required in order to make decisions on behaviour, either in terms of natural movements or control of movements due to a construction activity, yet to take place. Surveying methods are therefore used to monitor the magnitude and rate of horizontal and vertical deformations of structures, the ground surface and accessible parts of subsurface instrumentation; that is, to relate measurements of other instruments to a reference datum which cannot be tied in directly. Future courses of action need to be determined prior to commencing monitoring, to ensure that an appropriate system is specified. The data required need to be considered in terms of type, location and frequency, and use appropriate instrumentation for its collection. Part of the procedure is to determine the accuracies required and then the instruments capable of achieving those. These requirements need to be determined at an early stage.

It is necessary to determine what critical conditions are to be monitored on the project. These may include items such as:

(1) settlement
(2) tilt
(3) horizontal movement
(4) strain
(5) stress.

From these decisions will fall out appropriate monitoring methods. It is also necessary to consider what the required accuracies, frequencies and likely durations of the monitoring are. Any other project-specific requirements also need to be identified at this stage.

5.1.1 Global considerations

5.1.1.1 Absolute v relative measurements

It is important to determine whether the absolute or relative position of a monitored point is required. Absolute positions will generally require a greater effort to collect. The means of providing appropriate base readings against which those future movements will be compared needs to be considered. For example, a structure which moves bodily, say an overall settlement of 20 mm, is not normally of concern. It is differential movement which generally causes damage. In this particular situation, the utilities entering the building, being exposed to that movement, may be the major consideration, not the structure.

5.1.1.2 Real-time monitoring

When monitoring is required at a frequency exceeding that possible (or practical) by manual survey methods an automatic data collection system will be required. In these circumstances it may well be necessary to provide a 'real-time' definition for the project. An initial project stance is often that real-time monitoring is necessary because the equipment is capable of it. This approach will produce a very large quantity of data with its attendant processing and storage problems. It is more manageable to operate such a system taking readings at less frequent intervals, say hourly, and escalate frequencies in those areas of particular interest when necessary to allow detailed evaluations to be carried out.

The response time required between data collection and availability for use by the monitoring engineers will equate to a contract definition for 'real-time' monitoring. A strict definition of real time is virtually instantaneous availability of results, but survey type monitoring will not normally provide on a variety of points because there is a finite time for an instrument to physically move from one sight to another then return to collect those points unavailable during the first sweep.

5.1.1.3 Advantages/limitations of surveying (remote data capture) v electrolevel and other electronic specialist monitoring instruments

Advantages

- Minimises wiring requirements.
- Providing a line of sight is available, it is easier to check back to a point of reference outside a zone of influence for absolute positioning.
- Greater flexibility in adding extra monitoring points, after initial installation has been completed.

Limitations

- The mechanical nature of the mechanisms give a finite time to take a series of readings to different points, rather than virtually instantaneous.
- An optical line of sight is required; this includes during atmospheric conditions which may occur, such as fog.

5.1.1.4 System purposes

Primary system: If there is only one monitoring system on a structure it is, by definition, the primary system. This is the system on which main decision-making will be made. Any automatic electronic systems, whether survey or otherwise, needs a secondary manual fallback system for use in the event of malfunction of the primary system.

Secondary system: A monitoring system required to provide the back up and check facility to the primary system. It is often a manual check system on an automated system.

Strategic monitoring: A limited extent monitoring for structures which are predicted to be unaffected by the construction works but where it could be useful to demonstrate, at a later date, that no movement has taken place. Generally the only readings taken are those to set a base value and those after the construction operation. These readings could be considered as protection against spurious claims.

5.1.1.5 Economics (Manual v robotic systems)

Any monitoring must be appropriate to the location proposed. For example in a railway tunnel, if no access is available at the frequency readings are required, a robotic system is appropriate. However, a robotic station may not be appropriate if the area to be monitored changes quickly. Restrictions on safe access manual survey may also be a consideration.

There will usually be a frequency of readings at which a manual system is cheaper than robotic despite the higher manpower data collection costs. This point will depend on local economics, but the cost of a secondary system needs to be included in any such comparison.

Cost
Manual: Installation + cost/survey + processing time + reporting = total
Automatic: Installation + operation + processing + reporting + mainte-
           nance + removal + reduced frequency manual surveying
           and associated installations = total

5.1.2 General principles

Monitoring implies a more rigorous control than with normal surveying. Repeatability of readings is key and small differences can give advance notice of larger movements or of trend development, so they need to be right.

It is important for one person to be responsible for all aspects of instrumentation and have the time to co-ordinate all the necessary activities. Otherwise, if sufficient care is not taken through the installation, reading and interpretation processes, the results can become impossible to interpret. It is advisable to try and keep the same personnel on each job and a well-trained survey team aware of accuracies required is essential.

The principles involved in surveying monitoring can be split, typically, into the following categories.

5.1.2.1 Physical constraints

- Surveying instruments are generally used externally and rely on a line of sight, which tram wires, lighting columns, etc., may impede (see Figure 5.2).
- On a narrow street, acute angles will impact on accurate total station operation and discernment of targets along a façade, that is the very small angle to turn through may cause problems discerning between targets. With an automatic instrument where a large number of points are located in a small area it is possible that many targets will be seen at any one time. If one target is obscured, a brighter target adjacent to the actual target could be measured.
- Wind and vibration affecting instruments to give collimation errors.
- Beware of diffraction effects.

5.1.2.2 Readings

Standard surveying good practice should be followed regardless of whether automatic or manual readings are taken. For example:

- Take readings on both faces on a total station
- Check control on a regular basis
- Carry out observations at the same time each day or same day of the week.

Building Facade

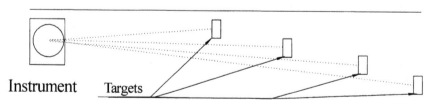

Instrument   Targets

Figure 5.2 Acute angle sighting problem.

5.1.2.3 Equipment

- Know the potential accuracies of the proposed system and ensure it meets requirements.
- Provide an element of redundancy within the monitoring system.
- If taking manual readings allow the instrument and tripod to acclimatise and try to protect them from direct sunlight throughout the surveying operation.

5.1.2.4 Frequency

The frequency of readings is often controlled by the rate at which the parameters being monitored are expected to change and the time anticipated before trigger levels would be reached. The readings must be frequent enough to identify significant changes in the behaviour of the structure and to give sufficient warning of an impending trigger level to allow required mitigation actions to be implemented in good time.

These phases of data collection can generally be split into:

- Before: Generally to build up a record of background readings to determine natural movements prior to commencement of construction.
- During: Frequency to match appropriate remedial measures which may be required during construction.
- After: Readings to demonstrate that construction-related movements have either ceased or reduced to a level that movement is trending out.

5.1.2.5 Base

It is important to have a valid base to compare construction readings against and to take sufficient base readings to take into account climatic and other environmental effects. Twelve months is a minimum period in order to determine annual climatic changes with any certainty, but it is unusual for a client to have a timescale of that length available in their programme. Three months is much more common and decisions will often need to be made later in the contract as to whether movements are construction-related or naturally occurring which could have been determined more easily if greater historic data were available.

A base can either be a single reading agreed by the parties as representative or as an arithmetic mean of a number of independent surveys. If there are considerable natural cyclic movements it may need a banding, either side of that mean, to be incorporated for analysis purposes. Regular readings should be recorded after installation of instruments, prior to any major construction works, in order to establish the range of background values due (for example) to tidal or temperature variations.

Levelling around the damp proof course to get existing settlements on older structures can prove a useful expedient, but will not obtain a meaningful estimate of existing differential settlement to an accuracy any better than the original construction. Structural movement limits are usually tied to the existing position of the structure, rather than its starting point.

5.1.3 Other considerations

5.1.3.1 General

- Vandalism is unfortunately a consideration at many locations. Monitoring points, if susceptible to deliberate damage, need to be located so that unauthorised access is difficult. For example, monitoring prisms fastened to a railway overbridge were accessed by vandals using timbers to extend their reach from an adjacent footbridge. The solution was to place a protective baffle above the prisms that did not prevent sighting down the side of the bridge from the instrument position.
- Positions of monitoring points need to be determined such that the construction works do not mean an early loss and necessitate reinstatement. Continuity of readings at a location simplifies reporting and record-keeping considerably.
- Trees, branches and leaves, etc, block lines of sight and prevent readings being taken. These may need to be removed or reduced in height prior to monitoring commencing.
- In these litigious times, it is worth stating that there are people who will go out of their way to try and hurt themselves on monitoring points so make them as inobtrusive as possible and with no danger to the general public.
- Use system checks to confirm correct functioning of the installed system through all its components.

5.1.3.2 Remote data capture

- Rodents eating electrical cables can cause a breakdown in data transfer, therefore protect the cabling with either trunking or place beyond animal attack.
- If a remote data capture type of monitoring system is in an exposed location, consider the effects of lightning in terms of both protection and minimising the effects if it does strike.
- Consider the use of cellular modems and solar cells for robotic installations at locations remote from landline telephones and mains electricity. In this eventuality it may be cheaper to use cellular lines at the data reception end as well in terms of call costs. Use of a cellular

modem will reduce the data transmission rates below that of a standard telephone line and whether this is satisfactory must be considered.

5.2 Precise levelling

Several manufacturers currently offer precise levels capable of a standard deviation for 1 km of levelling using an invar staff of 0.4 mm or better. An invar staff is necessary to achieve the order of magnitudes of resolution required. This may not be possible on say electric railways, where glass fibre may be required, but recognition of the lower accuracy achievable should be considered.

Instrument: Ensure that regular 2 peg tests are performed both before and after each shift to confirm instrument calibration with pre-agreed acceptance limits backed up by regular servicing. The specification needs to include accuracies achievable.

Survey operation:

- Brace staves using ranging rods or similar to prevent movement during readings.
- Chainstaff need to be aware of the importance of the accuracies needed and the need to keep the staff vertical.
- Use permanent stable change points. Can use monitoring points as change points or install points for this purpose.

Route:

- Use sight lengths of maximum 25 m and minimise number of change points consistent with this.
- Avoid grazing sights.
- Keep backsights and foresights of near equal length for any one set up of the instrument.
- Keep route the same, useful for manual survey but auto processing essential to keep same route
- Close all level loops to allow checks on closure values.

5.2.1 Manual data collection

5.2.1.1 Manual precise level (optomechanical)

This is the type of precise level that has been in use for many years, capable of levelling with careful use to ±0.3 mm for 1 km of double run levelling, using a parallel-plate micrometer.

The Mansion House (see Figure 5.3) is the official residence of the Lord Mayor of London. On the Mansion House, which was affected by the Docklands Light Railway (City Extension) construction in 1989 to 1991, the primary requirements were to observe and monitor the detailed deformation response of the structure when subjected to the construction of the overrun tunnel and associated works. The instrumentation was also used to monitor settlement which was evaluated in the light of allowable deformation values agreed with the relevant authorities prior to construction.

The primary monitoring systems for vertical movements were electronic, electrolevels and water levels installed in the basement being read on an hourly basis. The secondary instrumentation consisted of precise levelling points around the perimeter of Mansion House, adjacent to the internal electrolevels and water levels, and on adjacent structures. Accuracies for the levelling were demonstrated to be in the order of ±0.8 mm.

5.2.1.2 Manual precise level (digital)

A development of the precise level is the digital instrument which by reading a bar-coded staff automates the manual data collection operation. The operation becomes one of electronic data capture where data are stored in the level and downloaded to a PC at the end of the survey. Accuracies achievable are similar to the manual precise level, but with the advantage of automatic reading and logging, resulting in a speedier survey.

Figure 5.3 The Mansion House, London.

Measurement staves with an irregular but systematic pattern of white and black stripes of varying widths (the bar codes) are registered and decoded by the digital level As a by-product of this correlation process within the instrument the distance to the staff is calculated, and can be available on return to the office to confirm sighting distances, but these are not millimetre accurate. Normally if visibility is good enough for optical sighting with human eye and telescope then a digital reading should be possible. It is not usually possible to read a bar-coded staff with a conventional level, but some staves are double-faced with a conventional scale on the back.

Advantages:

- Eliminates reading errors and reduces the number of booking errors, generally limiting these to misidentification of points.
- Removal of transcription errors from field book to the monitoring database.
- Shorter survey times. Experience has indicated that fieldwork can often be completed in less than half the time of an optomechanical level with better precision of results.
- Instrument can be set to take inverted staff readings.
- Instrument can be set to repeat readings to confirm values prior to acceptance.
- Removable recording module units allow transfer to the office of survey results, whilst the instrument can continue in use in the field with a fresh module.

Limitations:

- Lighting of the staves needs to be infrared, not fluorescent. This complicates the taking of readings in low-light situations.
- Incompatibility of bar codes between various manufacturers.
- Maximum range in the order of 100 m.
- Instruments can cost two to three times the manual alternatives.
- Problems occur when the staff is not illuminated evenly, part in shadow, or possibly uneven lighting at night.
- There is a potential problem when taking digital readings at the end of the staff, as due to the measurement correlation process accuracy of readings at the ends of staff can be reduced.

Heathrow Express Railway, London: 1850 individual precise levelling points were installed on the original project with readings downloaded into a spreadsheet, on completion of the day's survey. This looked like, and could be examined in the same way as a level book, but with the results of the previous survey and base readings also available. Predetermined routes, rigidly adhered to, were necessary to allow automatic processing.

Primary system used for:

- Political monitoring on structures owned by third parties.
- Structures the frequency at which readings did not requiring automatic data capture.
- Subsurface cables/pipelines: these were surveyed at frequencies up to 8-h intervals whilst in the tunnelling zone of influence.

Secondary system used for:

- Structures where electrolevels were the primary system, such as car parks and passenger terminals.
- Existing tunnels to take readings at the crown and either side of the track.

5.2.2 Automatic data collection, digital precise levelling

These instruments are manual digital instruments with a drive mechanism and control feature adapted to allow remote control and programmed operation to take readings to fixed bar-coded staves. Self-checking routines are built into the instrument and errors can be reported remotely. Generally each instrument collects its own data which are transmitted to a central point for processing. The instrument does not therefore know its own absolute position and therefore needs to compare back to at least one reference staff. It may be worth considering the installation of a robotic total station to a digital level which can give an absolute level and relative horizontal movements. The instrument is best located in an area sheltered from the weather and direct sunlight.

Advantages: Ease and accuracy of transfer of level data along a chain of instruments. For example there may be three instruments in a chain, the middle instrument using a backsight to a stave location, which is actually a foresight, or intermediate sight for the outer instrument.

Limitations:

- Staves are fairly large and therefore obtrusive (0.5 to 1 m in length) and in order to get sights parallel down a façade the staves will need to project further out, as they become more distant from the instrument.
- Infrared light (not fluorescent) is required in order to operate, this will require artificial (halogen type) floodlighting at night. This is likely to prove an annoyance to local residents making them unpopular in residential areas.

- A good level of cleanliness of the staves is required for readings to be possible.
- Ease of instrument replacement without having to resight every point again and identify.
- Check how the accuracy of the instrument/target will be affected down the chain, that is from an external instrument to an internal one. The accuracies required will probably dictate the number of instruments in the chain before external control is once again available.

Questions to raise with potential suppliers:

- Details of historic downtimes experienced on other projects to date.
- What are the power supply requirements? Using a UPS, how long would instrument/logger box operations continue without mains power?
- What is the range of self-levelling correction within the instrument and when it is reached, can this be notified automatically to the monitoring office? In the location proposed, will the instrument move beyond its own in-built capability to accommodate movement?
- Is there an in-built error seeking routine to determine instrument malfunctions? If so would it notify the monitoring office?

Motorised level response time: Take a level value being read by an inner instrument on a chain of several instruments. A similar calculation can be performed for a chain of robotic total stations. Currently ten readings could be taken by one instrument in approximately 5–6 min. On that basis an instrument can read approximately 100 readings/hour, but recovery time will need to be built into the regime to return to those points not picked up during the first sweep and retry taking the readings. For example:

For Instrument A

Take 37 readings in one circuit and two of those points will be common with instrument B. Therefore 22 minutes to carry out a full sweep.

Readings taken	0100–0122
Time to pickup points that were obscured during first sweep (notional)	0123–0124
Database dials into logger and downloads	0125–0126
Validate and import into database	0127–0129
Output available	0130

For Instrument B

Fifty-eight readings to be taken in one circuit, therefore 35 min to carry out.

Readings taken	0100–0135
Time to pickup points which were obscured during first sweep (notional)	0136–0139
Database dials into logger and downloads	0140–0141
Validate and import into database	0142–0143
Output available	0144

On the basis of the above the minimum time to collect a level is the cycle time of the instrument within the chain which has the greatest number of observations to make. This will become the definition of real time which the overall system is capable of achieving, but this time will vary due to:

- The measurement mode used (single or multiple readings), that is one measurement, or say, an average of six to a single monitoring point.
- The targeting mode used (direct or search).
- The angle that the level must turn through to target all the staves.

5.2.3 Precise levelling points

Precise levelling points are placed on structures to allow vertical movements to be measured (see Figure 5.4).

Monitoring points should be placed as low down as possible to avoid (or at least minimise) the thermal effects of the structure being monitored. Where two separate points are close together it will be necessary to provide a means of identification to avoid misattribution of the results. Materials need to be sufficiently hard wearing to avoid damage during a rigorous surveying offensive, without being brittle. Marine grade stainless steel or

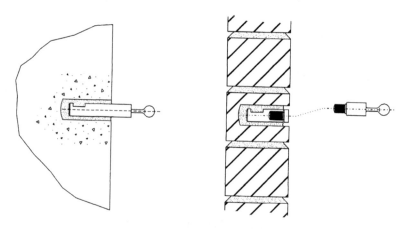

Figure 5.4 Fixed and removable BRE type precise levelling points.

brass is often used. Survey points installed near trafficked areas are prone to damage and disturbance.

Care needs to be taken to ensure that the monitoring point will react to the predicted movement in the manner anticipated. For example, if a precise levelling point is installed on a concrete slab it may in fact bridge the point of settlement and therefore not indicate the real movement. Similarly cladding will often not move in the same manner as the main structural members do.

5.2.3.1 Vertical plane (BRE type) to concrete/brickwork/steelwork

The special feature of the BRE socket (see Figure 5.5) is a loose-fitting screw thread, so those items screwed into it do not jam. There is a precise rebate into which a matching collar locates on the last turn of the inserted device, so that the device is located to a very high degree of precision, giving repeatable accuracy. When the point is to be used for levelling, the bung is removed and the levelling point screwed into position. To observe the monitoring point a precisely machined bolt is screwed into the socket for the duration of the measurement. This bolt has a rounded head which provides the definite high point to be monitored. During any fixing operation it is vital that any spillage of epoxy or mortar on the mating surface is removed immediately.

It is advisable to ensure that a common socket is used throughout the works to allow use of the same plug. This will make it easier for any longer term survey needs to be maintained. Where installation is into brickwork, the socket should for preference be installed to the brick, rather than the

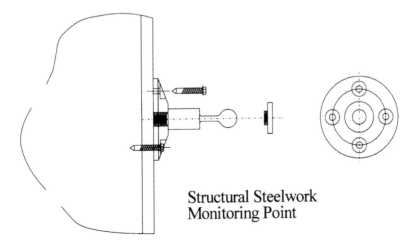

Structural Steelwork
Monitoring Point

Figure 5.5 Steelwork fixing detail.

mortar. This is because a socket is more likely to come loose from a mortar joint as building movement progresses. Aesthetics may however dictate that the point is made good and on this basis mortar is easier to match than brick (see Figure 5.6).

Perspex bungs are often supplied, but these can break and lock in the thread. An alternative is the use of a stainless steel threaded bung with dimples and a 'key' to allow authorised removal and prevent unauthorised access. A problem occurs where surveyors do not replace a plug, allowing dirt in. This means that it may not be possible to screw the levelling bolt in fully as the thread becomes partially obstructed with debris thus preventing the bolt from screwing to a repeatable position between surveys.

Plugs, which are notionally identical, need to be checked. It is worth having several produced and test each in turn on a point with same instrument set up to select those most consistent and avoid those levelling plugs with greatest geometrical variance. In locations where fixed precise levelling points are possible, these will give better repeatability.

5.2.3.2 Horizontal surfaces, concrete/tarmac

Domed head stainless steel rods are drilled and epoxied into the horizontal surface protruding slightly to give a high point to level to without providing a danger to users on foot or in road vehicles. Road nails, which are frequently used, are not suitable for precision work given their easy removal or displacement. Tops of kerbs are vulnerable to damage and more success may be achieved by installing in the road immediately adjacent to the kerb.

5.2.3.3 Subsurface monitoring points (earth anchors)

These are used for monitoring services below the ground. They are placed to the construction side of the service, to depth of the service as close, in plan, as it is possible to achieve without risking damage from the installation process itself. Settlement at the level of the service can then be

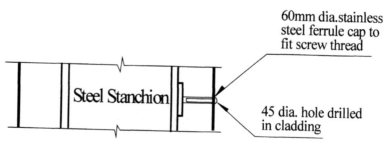

Figure 5.6 Cladding fixing detail.

determined from the ground surface. These anchors should, ideally, not be segmented to depth (see Figure 5.7).

5.2.3.4 Staves

Sometimes points are relatively easy to sight but physical access is limited, say a motorway overbridge central support, lying within the central reservation. It may be more practical to install a section of bar-coded staff permanently adhered to the structure involving only two traffic management exercises, installation and removal. These sections of staff can then be easily read at a comfortable distance without the necessity of a chain person visiting the point. For a remote data capture levelling system staves fastened to the building would be the normal situation.

5.2.3.5 Ground monitoring (no hard surface)

Sometimes ground surface monitoring is required and no suitable surface, tarmac or concrete, is available so a concrete block can be constructed and the monitoring point cast in or drilled and fixed. Sufficient time must be allowed for this to settle down and provision made to protect from construction operations (see Figure 5.8).

5.3 Co-ordinate survey

Co-ordinate survey is the measuring of points, within three dimensions, to a known grid. This usually necessitates the bringing in control, in three dimensions, from reference points remote from the effects of the works.

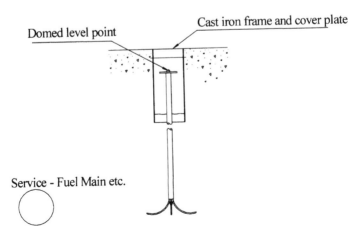

Figure 5.7 Sub-surface service monitoring.

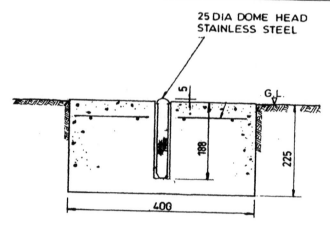

Figure 5.8 Soft surface monitoring.

- Useful to provide a monitoring point at a higher level if not possible to place a precise levelling point at a lower level.
- Resection: by observing five reference points in a good configuration it should be possible to achieve an accuracy of less than 1 mm in instrument position.
- Be careful when viewing stick-on reflective targets from oblique angles as distances will not necessarily be correct, but should be repeatable. When read from a different station an apparent significant change in position may be recorded.
- Use prisms appropriate to the instruments and accuracies required.

5.3.1 Manual

5.3.1.1 Total stations which need to be directed by a surveyor

Mansion House, London, secondary system: The primary system for horizontal movement was electrolevels, but an electronic co-ordination system made use of computer-linked theodolites and EDMs in three instrument set-ups to determine the position, in space, of reference pins on the structure. Two hundred of these reference pins (ceramic balls on metal pins) were fixed to key points on the masonry façade. These reference pins were located between electrolevels. Ten control stations were set up around Mansion House to provide a close reference frame for the spatial monitoring. Common pins between the instruments were then used to 'leap frog' around the building. A mixture of fixed pillars and tripod-mounted instruments were used. The processing involved linking each day's survey and importing into AutoCAD where horizontal displacements were plotted on an exaggerated basis on to a wire-frame model (see Figure 5.9).

Problems:

- The survey took 2–3 days to complete. It was determined that the top of the building moved up to 4 mm either side of centre during the day and therefore the surveys took too long to stitch together with an accuracy to indicate other than major trends.
- Traffic vibration and resulting effects on the instruments slowed the survey down.

Figure 5.9 Reference points, Mansion House.

- The instruments were connected by cable. It was not possible to avoid laying cable across trafficked roads and considerable cable damage resulted.
- The effectiveness of the spatial survey was severely limited by the interference of scaffolding works. The full extent of these scaffolding works was not known at installation.

Heathrow Express, cofferdam, secondary system: Following the tunnel collapse on the Heathrow Express project the solution was to construct a 60 m diameter 30 m deep cofferdam to encompass the collapsed ground. This was provided by the construction of interlocking piles, then excavating and constructing lining rings top down to provide an unencumbered working area. The objective of the instrumentation installed in and around the cofferdam was to monitor the response of the ground and the structure to construction (excavation) so that design assumptions could be validated and the expected behaviour of the design could be demonstrated.

Primary system: 3 m long bi-axial inclinometers installed to thirty-seven of the piles, spaced around the circumference.

Secondary system: Provision of holographic prisms fastened to the concrete inner lining at points corresponding to the ends of the inclinometers. 407 prisms on eleven linings (37×11). Survey was used to measure the deformed shape of the *in situ* concrete lining of the cofferdam and particularly its circularity (see Figure 5.10).

Four forced centring control stations were constructed at quadrant points around the cofferdam, SY1 to SY4. These were, however, within the zone of influence of the cofferdam construction. Although the measurement of relative movement between these stations was possible, the absolute co-ordinate values could only be monitored by reference to stable control outside the cofferdam and tunnelling zones of influence.

The only control visible from SY1–SY4 outside of the cofferdam were CTA 4, CTA 5, on multistorey car park 3 and CTA 6, CTA 8 on multistorey car park 1A, due to a close-boarded 3 m high fence. This control was not 'force centred', i.e. a tripod is needed to be erected each time they were used. These control stations were not established for planimetric monitoring and the co-ordinates were not of a precision necessary for this purpose. In addition, both car parks themselves fell within the tunnelling zone of influence (see Figure 5.11).

In order to avoid the problems of using the existing external control, a control network consisting of: CTA 4, 5, 6, 8, SY3 (the only forced centred control station then constructed) and four remote stations outside the zones of influence, were observed. The four remote stations consisted of datums 14 and 16 and two new stations 910 and 911. This control was computed on an arbitrary co-ordinate scheme then transformed to the local grid using

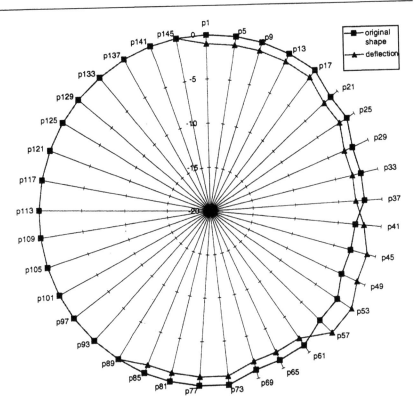

Figure 5.10 Heathrow cofferdam – deflection output format.

values of CTA 4, 5, 6, 8, but without applying a scale factor to avoid distorting the monitoring control.

The four remote stations; SD14, SD16, 910, 911: were not visible directly from the four forced centring stations SY1 to SY4; the external control was therefore transferred into the cofferdam using CTA5 as a temporary station. As this station lay within the zone of tunnel influence its co-ordinates needed to be recomputed each time it was used. Observations were taken to SD14, SD16, 910 and 911 and the position of CTA5 resected. Measurements were then observed to stations SY1–>SY4 as site conditions allowed. All observations to and from CTA5 were made without moving the tripod, therefore eliminating any centring error. Observations between stations SY1–>SY4 then completed the control network. Error analysis of these computations give an error ellipse for the instrument position at CTA 5 which indicated accuracies in the order of ±1 mm in eastings and northings as being achieved.

The instrument used for all control observations had a quoted S.D. of ±0.5 in for angle and ±1 mm + 1 ppm for distance measurement. These

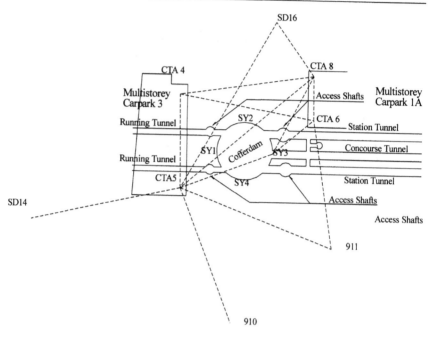

Figure 5.11 Heathrow cofferdam – co-ordinate survey network.

figures together with the results of the computation of the instrument position at CTA5 restricted the accuracy of the transfer of external control to stations SY1->SY4 to the order of ±2 mm in eastings and northings.

All monitoring points were observed from either the resected instrument position at CTA5 or SY1->SY4. Target tripods were centred over monitoring point using an optical plumb, which had a repeat accuracy of approximately ±1 mm. The absolute co-ordinate value of a monitoring point could therefore only be computed repeatedly to approximately ±3 mm.

The relative position of monitoring points from the forced centred stations was measured with a repeatability of approximately ±1 mm. Therefore to improve the accuracy of the monitoring the stability of stations SY1->SY4 needed be guaranteed and monitored.

To eliminate as many centring and transfer errors as possible when bringing in the external stable control on to the cofferdam site, it was necessary to establish a network of forced centred external control stations. From these stations it was possible to observe the force centred stations SY1->SY4 directly (see Figure 5.12).

National Grid Cable Bridge – primary system: At this location two tunnels were to be driven under existing 275 kV cables. The cables themselves could accommodate the predicted settlements, but there were two joint boxes located within the settlement trough which were less tolerant of this

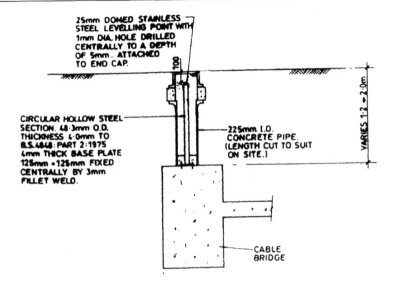

Figure 5.12 Monitoring of sub-surface structures.

movement. The solution was for the electricity transmission company to design and construct bridges in the ground to support the cables and joint boxes thereby protecting them from the predicted movements. The monitoring was to confirm that the settlement predictions were not exceeded and that the bridge performed satisfactorily. Readings were therefore taken to both points on the bridge and ground anchors to the surrounding ground. Base lines were set out extending beyond the predicted settlement trough, levels taken with manual digital levels and horizontal movement using a manual total station.

5.3.1.2 Non prism (reflectorless) total stations

There are total stations available which do not require prisms in order to take co-ordinate readings, but they cannot currently achieve the monitoring accuracies discussed elsewhere in this chapter. Accuracies typically 5 mm + 3 ppm are possible, but depend on reflectivity of the observed point and angle of incidence of the measuring beam. They are ideal where points to be measured are not accessible or not fitted with a reflector and not limited to particular frequency of prism locations.

5.3.2 Automatic (robotic)

There are now robotic total stations which can be controlled remotely, recognise and autoseek targets, then provide data automatically to a remote location using telephone or radio connection.

Instrument locations, to provide the monitoring coverage required, may need to be located within a predicted settlement trough and therefore cannot be fixed absolutely in space. In order to provide the absolute values required for control purposes, rather than relative values being taken by each instrument, it is necessary to tie the instruments back to fixed points. One instrument, at the end of the chain, will view fixed targets outside the settlement trough and can therefore determine its own, absolute position, in space and therefore the absolute position of the prisms that it observes. A proportion of these prisms will then be shared with the next instrument in the chain and the details of these points will enable the second instrument to establish its own absolute position and therefore the prisms it views. The rationale can then be passed down the line. In any specification involving a chain of instruments it will be necessary to specify an accuracy (to an absolute base) for both level and horizontal distance. This determination of position for the instrument should be performed both at the beginning and at the end of every measurement cycle.

Advantages:

- Will give three-dimensional co-ordinates, in space, for monitored points in absolute co-ordinates.
- Ideal for large numbers of points great distance apart.

Advantages (when used as a level, rather than a total station):

- Compared to a digital level relative movements of points (x, y) and absolute in z can be determined.
- Allows relative out of plumbs to be determined.

Limitations:

- Blocking of a prism face by dirt or damage will drag the instrument off-centre, necessitating a cleaning requirement.
- Telephone lines, may need two, one for control and one for download
- Limitations on number of instruments which one PC can control, each currently requires a serial port, therefore maximum of four per PC.
- Where there is a chain of instruments, say in an urban setting (restricted sight lines, along a tunnel trace) only some instruments will have access to points which can be considered as fixed. This information needs to be passed down the line. The accuracies required will dictate the number of instruments in the chain before external control is once again required.
- Self-seeking time/range and effect of a monitored point being missed and going back.

- Security, must be difficult to access to avoid theft/vandalism.
- Where remote data capture is used, a secondary system and manual download capability are still required.
- Very dependent on determination of instrument position accurately in space x, y, z. If the instrument is within the settlement trough its absolute position will alter and resection will need to be carried out every time a set of readings is taken. The accuracy of the monitoring readings is dependent on the accuracy of this resection. A minimum of five sights for these instruments will be necessary and may not be easy to achieve.
- It will be necessary to either carry a spare robotic instrument or have swift access to a replacement and these instruments are expensive.

Questions to raise with potential suppliers:

- Ease of instrument replacement. Does a new (or replacement) instrument need to be 'taught' its duties and where its sights are? Demonstration of accuracies quoted in a normal operating environment.
- Speed with which readings can be taken.
- Ease of taking control remotely to meet specific monitoring demands.
- What is the self-seeking range, usually in terms of an angle, and effect on a monitoring cycle of a missing point? Does it search for a set time before giving up and continuing its sweep or skip, finish its cycle, deliver the data to the logger then go back to find the point?
- How easy is transfer of co-ordinate data through a chain of instruments? This is likely to be a post-survey collection process run at the monitoring office.
- What is the standard deviation for the distance measurement?
- Accuracy of establishment of an absolute position achievable for the instrument coming from one point, three points, five points, by resection or other methods and the time to establish this.
- What effect will vibration of the types occurring in the proposed location have on reading accuracy and the ability of the instrument to take readings?
- Details of any self-checking routines built into instrument and how errors would be reported back by the system.
- Details of historic downtimes experienced on projects.
- How long would instrument/logger box operations continue without mains power using a UPS?
- Are any problems foreseen with frequencies of data collection required? For example the time for: (1) the instruments need to determine their own position in space (which may be dependent on values from an instrument further down the line). (2) The instruments need to collect their own particular data. (3) The data need to be transferred to the

monitoring office. What limit would this impose on the number of points an instrument could read within the time left?

- Range of self-levelling. Does it encompass predicted movement of its location and could a warning be transmitted to the monitoring office in this eventuality?
- Check how the accuracy of the instrument/target will be affected down the chain, that is from an external instrument to an internal one.

Example: Three high-pressure gas mains were to be horizontally, directional drilled at 5 m centres under five railway lines on an embankment. Measurements were required in plan and elevation to at least ±2 mm. Real-time reading was determined to be every 40 min on centreline of bore and every 2 h for the remainder. The fully automated total station with automatic target recognition, located on a 6 m high tower, was accessed by a free-standing scaffolding platform with observations made to miniprisms attached to the rails and output to a local laptop computer, with 24 h attendance by surveyors. No visible lighting at night was acceptable as this would draw unwelcome attention to the job and upset local residents.

The automated total station necessitated use of miniprisms rather than retroreflective targets, but this was offset by the decreased personnel costs. Seventy-two targets on specially fabricated brackets were attached to the rails. Eight control points were attached to various existing stable structures as equally spaced as possible with site constraints.

It was envisaged that the survey station would suffer movement from the effects of thermal expansion in the tower and therefore its position needed to be established at the start of each survey. Measurements were taken to all prisms 24 h/day for several days prior to construction, the eighty points taking approximately $1\frac{3}{4}$ h to observe.

Problems experienced:

- Vibration in the rails from train movements resulted in prisms twisting or falling off, despite thorough tightening of attachment bolts. This was overcome by applying locking-type adhesives to the bolt threads.
- The length of cable between total station and computer necessitated a reduced data transfer rate.
- Dust from construction site and dirt from the railway formed a film over some prisms rendering automatic target recognition inoperable. The quality of adjustments deteriorated if it was a control point that could not be read.
- The tower accentuated vibration from the trains and construction site plant and in some cases this exceeded the operating limits of the total station.

The system enabled vertical movement of the rails to be measured with millimetre accuracy 24 h/day. Trains continued to operate at full speed throughout the monitoring and construction process, which in turn contributed to a reduction in the overall cost and inconvenience of the pipe installation.

5.4 Global Positioning System (GPS)

GPS uses satellites in orbit to determine a position on the ground and is in itself a relatively complex subject. It is necessary to be able to receive these signals which may be shadowed by surrounding or overlaying structures. They do have limitations relating to satellite visibility and multipath errors, where signals are deflected by other objects nearby.

To date GPS is not sufficiently accurate to monitor building movements on a regular basis, not withstanding Professor Ashkenazi's important work on the Humber Bridge in 1996 (Ashkenazi *et al.*, 1996) which indicated that a resolution of ±1 mm horizontally and ±3 mm vertically were possible. Elsewhere specialist installations have been constructed for fixed locations where monitoring type accuracies can be obtained, but the cost of these currently precludes this approach on a more global basis (see Figure 5.13).

Another approach has been to use GPS as a means of locating another monitoring device. One example is as a means of locating a railway track-recording trolley, which itself is capable of tight dimensional control. This

Figure 5.13 Track monitoring vehicle.

is in order to provide a check for position in the event of wheel slip or incorrect initial placement of trolley.

The problem was to provide a three-dimensional record of track alignment (within 10 mm) relative to main project control, to act as trigger/alarm levels for any realignment actions. Limited possessions can make it difficult to create safe systems of work to allow for the conventional surveying systems of monitoring of the track without making the whole process very slow. This was overcome by utilising a conventional self-propelled track-recording trolley as the base machine.

The trolley itself permitted about 4 km of track to be monitored in a single, 1-h, possession. Using conventional surveying techniques to define absolute positioning would have significantly reduced the linear progression to approximately 300 m per hour. A GPS system was installed on a standard track monitoring trolley and GPS data measurements automatically taken at intervals. Two independent computers were carried on board: one to monitor the track recording vehicle normal functions and the other the GPS measurements. The two computer clocks were electronically synchronised before the start of each measurement run to allow combination of the two data sets in a post-processing exercise.

GPS is likely to be more frequently used in future, but needs to become substantially cheaper and more accurate for a distributed remote data capture system to become viable.

5.5 Other survey methods

5.5.1 Plumbing

5.5.1.1 Optical

Optical plummets are produced by a number of manufacturers and produce a vertical line of sight either up or down.

On Heathrow Express there were two car-park access towers containing banks of lifts that were adjacent, but separate, to the main car park structures and thus free to move. They had a high height to base ratio and were situated adjacent to large station tunnels and concourse tunnel excavations. Monitoring points were affixed to the bottom to mount instruments and gridded perspex targets permanently attached to the top. This was a secondary system to the horizontal electrolevels and precise levelling points at the base of the towers.

5.5.1.2 Spirit level

For determination of plumb of minor structures and facilities such as signals, the humble spirit level should not be disregarded in terms of a monitoring tool for measuring tilt. A 1 m level can with careful use be used to check plumb to within 0.5 mm/m, i.e. 1 in 2,000 which for minor structures or facilities that can be considered as moving bodily may suffice. It is vital in these circumstances to mark the monitored position to ensure repeatability and ensure movements in both axes are recorded.

5.5.1.3 Theodolite

A theodolite can be used if access is available beyond and in line with the structural entity to be plumbed plus an adequate target can be sighted and a means of measurement at the base determined.

5.5.2 Tape extensometers

Tape extensometers are used to determine changes in the distance between reference points anchored in walls or structures. Examples include the monitoring convergence of tunnel walls, deformations in underground openings, displacements of retaining structures, deep excavations, bridge reports and other concrete or steel structures (see Figure 5.14). Tape extensometer readings can be taken within underground tunnels affected by new works as a check on ovality. These can be augmented by precise level readings taken at the crown with a suspended staff and precise levelling points in the invert. Where existing tunnels are constructed from concrete linings,

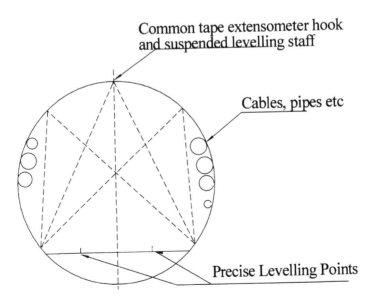

Figure 5.14 Typical orientation of tape extensometers in tunnels.

and it is often unacceptable to drill these, holes left from the original construction can be utilised to fix the extensometer eyes in.

Advantages:

- Compact and light
- Rugged
- Easy to use and maintain
- Initial alignment not critical, because hook and eyebolt can accommodate almost any tape angle

Limitations:

- Intended for relative measurements only
- No thoroughfare possible whilst readings being taken.

5.5.3 Datums/bench marks

Stable, reliable reference datums are required for all survey measurement of absolute deformations and decisions need to be made as to whether vertical stability, horizontal stability or both are required (see Figure 5.15). Nearby structures may perform a better datum function than any short-term construction-related items; for example, if deep-seated buildings on piles are available then these may provide a better datum than anything which could be installed as control. Old buildings with a basement may be a good choice whilst a new one may be heaving still from its construction.

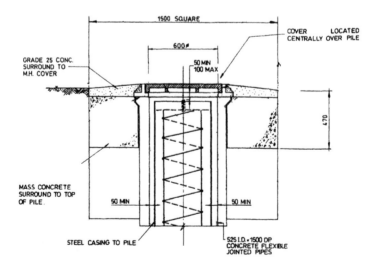

Figure 5.15 Typical isolation detail datum/surface.

Alternatively rising water tables may make the basement structures less suitable. If specific datums are being considered they should be installed to a depth below the seat of the vertical/horizontal movement. If existing benchmarks are being considered for use they may not have been designed to be isolated from surrounding ground movements. In some ground conditions it may be that only vertical stability can be obtained with any degree of certainty by going deep to solid underlying strata and that horizontal stability cannot be achieved (see Figures 5.16 and 5.17).

Datums should be located, wherever possible, outside the zone of influence of the construction works. If datums need to be located inside the area of influence then more frequent checks will need to be undertaken to ensure that continuity of data is available. Consider seasonal influences or regional movements, for example, due to ground water changes.

On Heathrow Express 22, datums were constructed along the route. These were concrete cast *in situ* piles installed through the Thames Gravels approximately 3.6 m into the underlying London clay, outside the zone of influence of tunnelling. These piles were isolated from the surrounding ground at the surface by being placed within a concrete pipe, providing a clear annulus. The covers incorporated small central inspection hatches allowing easy access to the monitoring point by the surveyors. These dat-

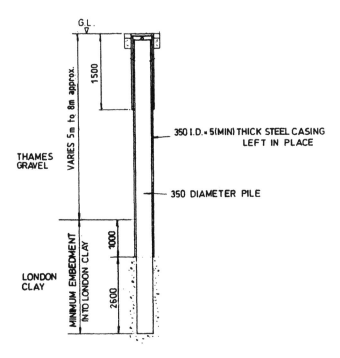

Figure 5.16 Pile datum details.

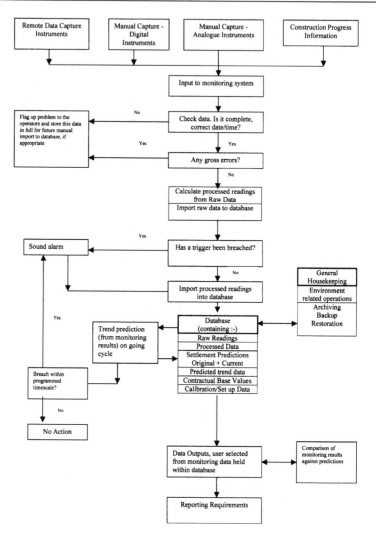

Figure 5.17 Data Processing.

ums were used for project-wide control and not just monitoring. Levels between them were checked weekly and co-ordinates fortnightly during tunnelling.

5.6 Survey data management system

Successful monitoring data collection and management are of vital importance in many construction projects. A breakdown of the process can result in programme delays, additional costs and possibly structural failures. For successful collection and management, it is necessary to plan and instigate an appropriate system at commencement of the work and to manage that system, in all aspects, throughout the construction period, including both background readings to determine existing trends and close out to confirm that movements have ceased.

Rapid retrieval and delivery of processed data in both hard and soft form, in appropriate formats, are fundamental to managing risks to projects. The late collection and assessment of data has brought works to a halt in the past.

5.6.1 Identifiers

Any surveying data management system needs unique identifying numbers and is usually incorporated within a larger instrumentation system. The lowest common denominator requirements need to be used, for example where early digital levels are used they could only take numbers 0 to 9, so letters were not used. It is important to establish an appropriate logical numbering system. By way of example on Heathrow Express, each surveying point had a unique six-digit identifier, AABBCC where: AA, the last two digits of the drawing number on which the monitoring point appeared; BB, the survey line on the drawing, commencing at 01 on each drawing; CC, position of the point within that line.

5.6.2 Processing

Consideration needs to be given as to how the data collected will be processed, output and reviewed.

The users will select the required information from the processed data tables. It is useful for the front end to be sufficiently flexible so that users can make their own reporting selections and not be tied rigidly to pre-defined formats. A form of selection of the point of interest needs to be determined and then the user can select from those points for which they wish to examine the data. After selection of the points is completed the options regarding date/time, etc., will need to be made available.

Consideration should be given as to which data will be supplied for a particular date/time. Take as an example a group of precise levelling points that are actually surveyed on different days within the group (that is the surveying grouping does not correspond to the current user selection). Where data for a particular point are not available for the date/time selected, which value should the system select/return: (1) The last reading taken prior to the date/time requested; (2) the next reading after the date/time requested; (3) a proportioned value between (1) and (2) for the date/time requested.

There is no correct answer for all eventualities. (3) is generally likely to be the preferred option except where a significant short-term movement takes place, when it could be very misleading.

The software needs to encompass data collection, storage, transmission, verification and processing manual, hand-written + manual, digital + automatic, digital data and the following points should be considered:

- Need to take temperature measurements as well to determine likely thermal effects on the monitored buildings.
- Trend prediction: it is often important to know when, based on current rate of movement, a limit will be breached. It is better for a simple approach to be used, possibly based on a straight line, best fit through a number of the most recent readings projected to meet a point in the future. It is likely to be a fine balance between one reading tipping the balance and an insufficient reaction.
- Frequency of reporting needs to be considered and flexibility of output, both graphical and tabular.
- Amber and red trigger levels need to be incorporated and either warning on input for manual collection or automatic during processing for RDC instruments. An RDC system can be programmed to set off a pager or deliver a message to a mobile phone. Provision should be made for escalating the call if the mobile phone does not respond. It will prove easier to administer if the on-call engineer receives the monitoring phone rather than rotating the numbers within the notification program. It may be useful for the on-call engineer to be able to dial into the monitoring office using a laptop and run the monitoring software remotely. The optimum method of achieving this is using the laptop merely to mirror the server computer in the office, thus reducing the amount of actual data transmission down the telephone line.
- It is inevitable that there will be a certain amount of noise within monitoring schemes and that should be appreciated so as not to ring alarm bells unnecessarily.
- It is important to store both raw and processed data so that re-processing data can be provided in future if problems with the initial processing are discovered at a later stage.
- If data are required to be usable beyond the end of the project and into the future the storage medium must be selected to ensure future access. As an example in 1993, much data were stored on $5\frac{1}{4}$" floppy disks, how many offices still have the ability to read data from this source? That is merely taking the most common of storage medium, not other less commonly used which are no longer available.
- Size the database to hold the maximum number of results and run tests regarding access times with dummy data to confirm response times.

5.6.3 Instrument control, remote data capture

Once set up these installations should be the easiest to administer, but the following should be considered:

- Ability to read battery level remotely by the monitoring office team
- Does any reprogramming of the instrument need to await a gap in monitoring or an additional telephone line?
- Are manual downloads possible in the event of telephone line loss?
- Is the control program intuitive and are changes in logging frequencies, etc., easily implemented?
- Are simultaneous data collection and downloading to the monitoring office possible or advisable?

5.6.4 Reporting

For interpretation purposes, there is a need to see graphs which show movement against time, to determine if trends are developing. Graphical output is easier to assimilate, but the tabular information needs to be readily available when closer examination of actual values is required. One useful plot is along a line with time, that is, y-axis is mm of movement and x-axis is length along the line. Results are plotted as lines for a certain date, enabling several dates to plotted together. If lines are parallel, body movement has occurred.

Facilities for the comparison of monitoring results with predictions should be available. User selected combinations need to be stored for future use, therefore a means of satisfactorily identifying them needs to be determined.

5.7 Conclusions

The advantages are such that precise survey techniques will continue to be included in any instrumentation scheme for the foreseeable future. Where other systems may not function due to power failures, cable damage or electrical storms, precise survey data are always on hand as a backup. Recent developments in robotic total stations provide a real alternative to electrolevels for an automatically operating installation.

Properly installed and used instrumentation can assist engineers to make correct decisions that can save lives, time and money. Instruments do fail and that is why experienced engineers specify extra instruments to take this into account. With the right attitude and guided with the proper training and knowledge, the likelihood of things going wrong will be minimised. This chapter will be judged a success if others can learn and avoid some of the potential pitfalls outlined above.

Manual surveying type monitoring activities can be a very repetitive but highly labour-intensive task. The utmost care must be taken to ensure that established surveying practices are followed and that detrimental short cuts are not taken. Stringent quality controls must be implemented and maintained throughout the data collection and processing and data presented in a timely fashion to allow decisions to be made in an optimum fashion.

Finally properly managed monitoring is an investment that has the potential to offer substantial cost savings in the future by providing an early warning of undesirable movements.

5.8 Glossary

Amber trigger limit	A value set at a proportion of the main limit imposed, this provides an awareness trigger that allows time to implement actions in the event of movements taking place. See also red trigger limit.
Base readings	Data collected prior to the commencement of construction work where underlying trends are established.
Bi-axial inclinometer	An instrument which measures movements perpendicular to its main axis, often used to monitor horizontal sub-surface movement.
BRE	Building Research Establishment
Calibration	Calibration is required where instrument records data in the form of, say, voltages and requires calibration for conversion to movement in millimetres and is normally carried out in a laboratory.
Datalogger	Connected equipment to record data from electrical instrumentation and may comprise a multiplexer, A/D converter, amplifier, volt meter and storage memory.
Electrolevel	Electronic tilt meter, comprising a vial partially filled with electrolytic fluid, penetrated by electrical conductors. A change in tilt is measured by a change in resistance or capacitance across the conductors. Given a known length the angle through which the instrument has moved allows a displacement to be calculated.
GPS	Global Positioning System
Red trigger limit	The value at which an action needs to be taken. See also amber trigger limit.
Real time	For automatic type surveying instrumentation, hourly readings under normal operations. Maximum points an instrument can read in a time currently forty readings an hour needs a contingency to allow recovery for returning to sight points which could not be found on initial search for RDC. A direct reading could however

be taken off one point continuously at the expense of other points.

Remote data capture
: An instrument capable of taking results capable of taking results remotely in a digital form and transmitting these data to an office not immediately adjacent to the monitored location.

Return of data time
: The time interval between collection of the last piece of data by an instrument from a regular monitoring sweep and provision of that data, in absolute meaningful terms to the monitoring team.

Settlement trough
: Ground movements associated with subsurface works are manifest at the surface in a trough form in front of and to the side of the advancing tunnel face. This is termed the settlement trough.

System check
: An *in situ* check to confirm that the system, in its entirety, is functioning correctly and providing data which are correct. For example, for a robotic total station this would include:

- the collection of the reading and confirmation that the value is correct and correctly identified
- processing for onward transmission
- correctly storing and sending or collection from the datalogger to the monitoring office database.
- correct placement and retrieval from that database.

Note. This is not a replacement for laboratory calibration of individual instruments.

Total station
: A combination of electronic theodolite with an electronic distance meter to measure distance.

UPS
: Uninterruptible power supply

Zone of influence
: The area at a particular level affected by the works, for example for a tunnel constructed at depth the zone of influence just above tunnel level will be significantly less than that at the surface.

5.9 Acknowledgements

The author would like to thank his colleagues for their help and support during the preparation of this chapter. Any opinions expressed in this chapter are those of the author and not necessarily those of his employer.

References

Forbes, Bassett and Latham. 'Monitoring and interpretation of movement of the Mansion House due to tunnelling', *Proc. Instn. Civ. Engineers Geotechnical Proc*, April 1994.

Cook, D.K. 'Settlement monitoring on the Heathrow Express – Engineering Surveying Showcase', October 1996

BRE Digest 343. Simple measuring and monitoring of movement in low-rise buildings: part 1: cracks. Building Research Establishment, Garston, UK.

BRE Digest 344. Simple measuring and monitoring of movement in low-rise buildings: part 2: settlement, heave and out of plumb. Building Research Establishment, Garston, UK.

BRE Digest 386. Monitoring building and ground movement by precise levelling. Building Research Establishment, Garston, UK.

BRE CP26/73. 'Techniques and equipment using the surveyors level for accurate measurement of building movement', J.E. Cheney. Building Research Establishment, Garston, UK.

Cook D.K. and Akbar S. *Data Collection and Management*. FEGM99, Singapore.

Geotechnical Instrumentation for Monitoring Field Performance, Wiley-Interscience, J. Dunnicliff, 1993.

Cook, D.K. (1996) 'Settlement monitoring on the Heathrow Express', ICES Biennial Conference.

Ashkenazi, Dodson, Moore and Roberts. (1996) 'Real time OTF GPS monitoring of the Humber Bridge', *Surveying World*, 4, 4.

Uren, J. and Price, W.F. (1994) *Surveying for Engineers*, MacMillan Press.

Chapter 6

Practical bridge examples

Klaus Steffens

6.1 Introduction

The motives, methods, prerequisites, conditions and experience in the application of experimental assessments of load-bearing verification have been dealt with generally in Chapter 7. In this chapter, the special aspects of *in situ* load tests on solid road bridges are highlighted using examples.

The complexity of the task definition, the high level of risk, the heavy test loads and varying load positions demand a high degree of experience and care in the planning and execution of load tests on solid bridges.

6.2 Requirement

The methods of the experimental assessment of load-bearing capacity on solid bridges was further developed in an extensive co-operative research project EXTRA II (Steffens, 1999) between the University of Applied Science Bremen (overall responsibility), the Technical University Dresden, the Bauhaus University Weimar and the University for Engineering, Economics and Culture, Leipzig. To clarify the need for the application, data from a total of 361 bridges in Mecklenburg-Western Pommerania (northeastern Germany) was assessed in a statistical survey. It became apparent that about 80% of these bridges are of reinforced steel construction, have spans of 1–18 m, exhibit defects with regard to the condition of maintenance or load-bearing capacity and can be classified in the Bridge Class 30/30 according to DIN 1072 (1985) and are therefore in principle suitable for *in situ* load tests.

6.3 Loading technique

6.3.1 Preliminary remarks

The sensors for recording the reactions of the structure under external test loads can be compared with those used in general civil engineering and are supplemented, where required, by acoustic emission analysis and dynamic measurements for system identification.

The following versions of the loading techniques were used for producing external test loads, ext F, on solid bridges in the pilot objects of the EXTRA II research project.

- Lorry/mobile crane for crossing bridges (service loads)
- Stationary masses for test load activation
- Loading frame from the EXTRA I programme

The various versions are characterised by their main parameters:

- Production of service loads/test limit loads
- Load sizes
- Load control, load measurement
- Geometrical bridge shape, support conditions, spans
- Roadway sealing
- Set-up time and procurement costs
- Possibility of restricting bridge use during the tests

In the following, the various versions are compared with one another and from the experience gained, requirements for the development of a loading vehicle are formulated.

6.3.2 Lorry/mobile crane

On account of the lack of a fail-safe mechanism due to unannounced component failure (risk of damage and collapse), mobile test loads, for example in the form of heavy duty vehicles, are only suitable for producing uncontrolled service loads, which are proven by computation and in the most favourable cases also slightly above this level. Otherwise, with regard to the main parameters they can be used very much to advantage. Here, the application is the bridge over the canal at Ludwigslust in north-eastern Germany (see sections 6.4.1 and 6.4.2).

6.3.3 Stationary masses

With this method masses are placed, secure from collapse, such that the test loads can be produced, continuously adjustable, with a hydraulic loading system operating against the weight of the masses. This version has been successfully applied with extreme boundary conditions (see sections 6.4.3 and 6.4.4), but it is generally impractical for standard bridges of Bridge Class 30/30 with spans over 15 m.

An interesting application example here is the experimental assessment

G = ◧◪◫ t
L = ◧◪◫m

Figure 6.1 Vehicle travelling position. *Source:* Eggers Fahrzeugbau, 28804 Stuhr, Germany.

of load-bearing capacity on the prestressed concrete bridge over the Main-Danube Canal at Baiersdorf (Bavaria), undertaken within the scope of the project EXTRA I (Busch *et al.*, 1995; Steffens *et al.*, 1995; IWF, 1997).

6.3.4 Loading frame

In the EXTRA I research project steel loading frames where developed as a kit system, initially for building construction. Due to their high load limit of 750 kN per frame, these are also suitable for bridges, provided reaction anchorage of the frame against the superstructure or the abutment or pillar is possible. The main disadvantage of this method lies in the necessity for drilling holes in the superstructure sealing for the passage of the tension rods and their closure after the conclusion of the examination, as well as the relatively high expense in terms of time and cost for transport, setting up and dismantling (see section 6.4.5, 6.4.6 and 6.4.7).

6.3.5 Loading vehicle (special vehicle for bridge tests)

Extensive experience with solid bridges (about forty-five assessments of load-bearing capacity) obtained by the co-operating parties show that the development and testing of a special loading vehicle for standard bridge tests is required. With the availability of such a vehicle it is expected that the examination costs and the closure periods for road and pedestrian bridges could be significantly reduced (Figures 6.1 and 6.2).

Specification (extract):

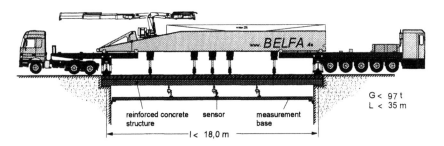

reinforced concrete structure   sensor   measurement base

l < 18,0 m

G < 97 t
L < 35 m

Figure 6.2 Working position. *Source:* Eggers Fahrzeugbau, 28804 Stuhr, Germany.

- Two vehicles of the same type
- Alternative operating principle according to sections 3.2 and 3.3
- Vehicle masses should enable load tests for Bridge Classes up to 60/30
- The vehicles must still be able to cross road bridges of Bridge Class 12 with spans of up to 12 m in order to reach their deployment location
- Dimensions and axle loads must conform as far as possible to the road traffic registration directive (StVZO) in order to ensure operation as a normal vehicle.

6.4 Application examples

6.4.1 Bridge of the Holy Spirit in Hamburg

The stone arched bridge built in about 1880 has three spans with a total length of approximately 34 m. After foundation work in front of an abutment, a horizontal offset in the imposts had set in. After injection of the masonry cracks the success in the restoration and the serviceability of the bridge was to be verified (Figure 6.3).

The service-load test took place with two mobile cranes, each of 36 t mass, moving in parallel. A corresponding position measurement was used

Figure 6.3 View of the bridge.

as the independent variable for the on-line display of the deformation effects (Figures 6.4, 6.5 and 6.6).

6.4.2 Canal bridge at Ludwigslust (north-eastern Germany)

Traffic development necessitated increasing the bridge classification from 16/16 according to DIN 1072 (1985) to 30/30. The poorly defined load-bearing system had to be identified by two mobile cranes crossing with the service load to obtain system data for the finite element analysis (Figures 6.7 and 6.8).

6.4.3 Pedestrian bridges at Loxstedt (Bremen)

Bridge owners have a duty to maintain traffic safety over road and pedestrian bridges, to keep bridge logs and to classify the relevant structure under bridge classification according to DIN 1072. Sometimes, as a supplement to computational verification, an experimental assessment of load-bearing capacity is required in cases of moderate damage affecting static properties or where the service loads are to be increased. In the case of the Loxstedt bridges, the variable test load was produced by moving a mass (36 t mobile crane) on a point-supported, fail-safe load-bearing frame and

Figure 6.4 Mobile cranes as test load.

Figure 6.5 18 m measurement base.

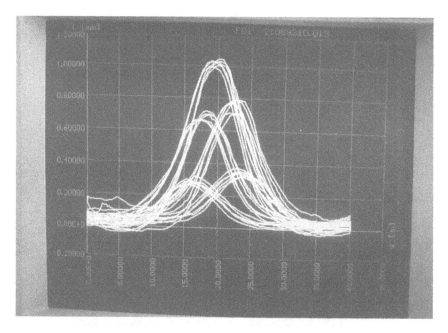

Figure 6.6 Monitor image of the vertical arch displacement at the quarter points in the central span.

Figure 6.7 Production of the test load using two mobile cranes.

Figure 6.8 Inclination measurement at the point of support for the verification of the slab fixing in the abutment walls.

the continued usage for an increased bridge class was ensured (Figures 6.9, 6.10, 6.11 and 6.12).

6.4.4 Main-Danube canal bridge at Baiersdorf (Bavaria)

The bridge at Baiersdorf has a maximum span of 56 m. Due to corrosion of the prestressed steel, it was classified in Bridge Class 6 before the experimental examination and after conclusion of the evaluation of the test results, it was able to be approved again for the Bridge Class 16/16, (Figures 6.13, 6.14, 6.15 and 6.16).

Here, the mass of a ballast-loaded lighter which was let down on to the appropriately prepared bed of the canal was used as the fail-safe counterweight for accommodating the load reaction forces. The force was produced hydraulically.

6.4.5 Stepenitz bridge (northern Germany)

Along Federal Highway 105 between Lübeck and Wismar, the Stepenitzbrücke was widened and reinforced constructively in 1991 using a topping concrete slab. The bonding strength could not be verified by computation and had to be determined experimentally. Six loading frames from the

Figure 6.9 Mobile test equipment.

Figure 6.10 Reaction load using a mobile crane on a steel supporting frame.

Figure 6.11 Mobile crane in the working position.

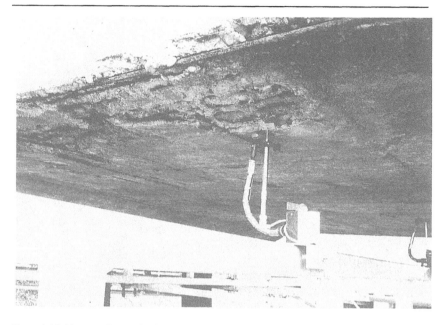

Figure 6.12 View underneath the damaged bridge slab.

Figure 6.13 Aerial view of the test site.

Figure 6.14 Bridge loading against the fail-safe mass of the ballast-loaded lighter.

Figure 6.15 Loading frame.

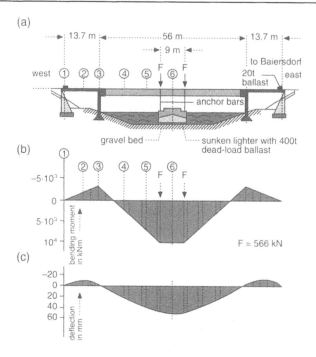

Figure 6.16 Longitudinal section of the Baiersdorf Bridge. (a) Loading arrangement with the LASH Lighter, lying on the canal bottom and loaded with ballast. The lighter acted as an abutment for the anchor bars used for loading. (b) Moment arising under the test load (times γ) referred to as a T-beam web. (c) Trace of deflection under the test load (times γ).

EXTRA system were used with the bridge closed on one side (Figures 6.17 and 6.18).

6.4.6 Road bridge at Ditzum (East Fresia)

An area of new building construction was to be linked via an 18 m long prestressed concrete bridge. Due to the lack of constructional documentation and an unknown degree of prestressing, the load-bearing capacity of the bridge had to be verified experimentally (Figures 6.19 and 6.20).

6.4.7 Wooden bridge at Güstrow (north eastern Germany)

The existing wooden bridge with nineteen spans each of 10 m width consists of four parallel nailed plank supports with a wooden roadway on wooden pile yokes. The wooden construction exhibited various types of damage, the severity of which could not be reliably estimated due to inac-

Figure 6.17 Test set-up.

Figure 6.18 Equipping of all three bridge spans with measurement bases and measurement transducers.

Figure 6.19 View of the bridge with loading frame.

Figure 6.20 Hydraulic production of the test load and electrical force measurement.

cessible sections and areas. Load tests were able to provide some clarification (Figures 6.21, 6.22, 6.23 and 6.24).

6.4.8 Weir bridge over the Weser at Drakenburg

A weir bridge that was being taken down, having spans from 11 m up to 42 m and parts up to 140 t in weight, was broken up for research purposes and brought by ship and road vehicle to the test department in Bremen. The approximately 1-year long examinations on reinforced concrete and prestressed concrete parts by the EXTRA research group (see section 6.2) brought significant, fundamental knowledge for carrying out experimental verifications of load-bearing capacity in solid bridge construction (Steffens, 1997) (Figures 6.25, 6.26, 6.27, 6.28, 6.29 and 6.30).

6.5 Summary

The projected service life of solid road bridges is about 80 years. Due to the damaging effects of the environment and increased traffic loads, it is in reality reduced to about 50 years. The ensuing need for reinvestment can only be covered taking into account financial resources and environmental concerns if the service lives of existing bridges are significantly extended.

Figure 6.21 Damaged nailed-plank main supports.

Figure 6.22 Trial loading with mobile loading frame.

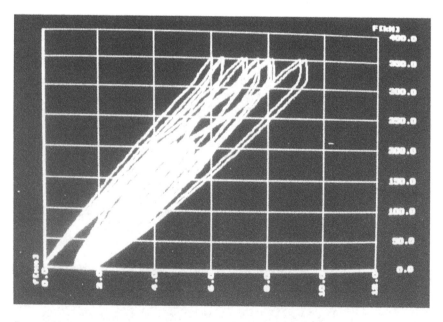

Figure 6.23 Reversible force/deformation curves with bridge spans intact.

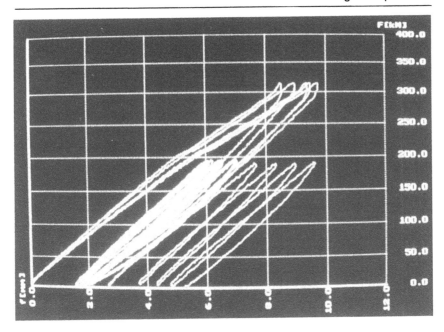

Figure 6.24 Irreversible force/deformation curves with bridge spans damaged.

Figure 6.25 Removal of the 42 m prestressed concrete beam (140 t) with the floating crane.

Figure 6.26 Road transport to the test department.

Figure 6.27 Aerial view of the test department in Bremen.

Figure 6.28 42 m prestressed concrete beam under final test load with 40 cm deflection.

Figure 6.29 Loading equipment for the dynamic tests.

Figure 6.30 Recycling the materials.

Load tests on solid bridges can make an important contribution here (Bucher *et al.*, 1997).

The technology transfer of experimental verification of load-bearing capacity (load tests and monitoring), as a supplement to computational assessment has been developed through to the application stage and will soon acquire the necessary acceptance.

References

Bucher, C., Ehmann, R., Opitz, H., Quade, J., Schwesinger, P. and Steffens, K. (1997) 'EXTRA II – Pilot object, Weser weir bridge, Drakenburg. Experimental stability assessment of solid bridges (in German)', Bautechnik 74, issue 5, 301–319.

Busch, E., Ehmann, R. and Steffens, K. (1995) 'Prestressed bridge over the Main-Danube canal at Baiersdorf, hybrid stability and serviceability assessment (in German)', Bautechnik 72, issue 3, 152–162.

DIN 1072. (1985) Straßen- und Wegebrücken (road and pedestrian bridges): Design load, Beuth Verlag Berlin.

Experimental assessment of load-bearing capacity – prestressed bridge over the Main-Danube canal at Baiersdorf, Film C 1990, 1997, 14 min. (in German), Göttingen: IWF GmbH.

Steffens, K., Tschötschel, M., Ehmann, R., Opitz, H., Quade, J. and Schwesinger, P. (1995) 'Measurement and testing techniques used in verifying the load-bearing capacity of a bridge over the Main-Danube canal', *Reports in Applied Measurement* 9(2), 25–31.

Steffens, K. (1997) 'Experimental assessment of load-bearing capacity of bridges *in situ* for the retention of capital investment and for the reduction of environmental pollution', Co-operative Research Project 01-RA 9601/6 EXTRA II (in German), Intermediate Report, Drakenburg: University of Applied Science Bremen.

Steffens, K. (ed.) (1999) Experimental assessment of load-bearing capacity of bridges *in situ* for the retention of capital investment and for the reduction of environmental pollution, Final Report of the Co-operative Research Project EXTRA II (in German); University of Applied Science Bremen.

Chapter 7

Practical building examples

Klaus Steffens

7.1 Introduction

In terms of planning and type of construction, structures are often designed for special requirements and a limited period of use. In the age of sparse resources, frequent changes of use, environmental damage and the consequences of inadequate building maintenance, questions arise with existing structures regarding the change of use, building modifications and the remaining service life, as the basis for decisions about investment. A prerequisite for the retention of the capital represented by existing structures is the verification of their load-bearing capacity and serviceability. The usual computational analysis assumes that, apart from the geometry, support and loading, all the major material properties and state characteristics are known and that behaviour under load can be described realistically by mathematical methods.

The reality of construction shows though that in many cases one or more prerequisites for the computational verification of load-bearing capacity are not known or can only be determined with uncertainty. The reasons for this might be:

- unsatisfactory or missing static documentation
- defects in the building construction
- material deterioration
- inadequately defined load path, inappropriate model formulation
- changed (increased) requirements due to modifications and changes of usage.

In such cases it is sometimes worth carrying out a loading test *in situ* on the existing structure. A prerequisite here though is that this is carried out without causing damage which might impair the load-bearing capacity or the durability of the object. The application of a reliable, economic and rapidly deployable, mobile on-site measurement system is required.

In this chapter and in Chapter 6 (Bridges) the fundamentals of these methods and their technology are described, and their advantageous application is illustrated using practical examples.

7.2 Methods

The method of experimental verification can be most simply described as shown in Figure 7.1 (Anon, 1999).

During the loading test, an existing structural component with an (unknown) effective load-bearing resistance eff R_U is stressed, subsequent to a prior assessment of the actual condition, with additional applied loads and their effects on the structure (e.g. deformations) being measured. At the start of the test the sustained loads G_1 are already present in their actual magnitudes. During the load test, additional external loads ext F_{target} are applied which are given in Guidelines for Load Test on Solid Structure (1999). The sum of G_1, G_{dj} and Q_d, which represents the test target load F_{target}, must not be greater than the test limit load F_{lim} for which the criteria according to the guidelines are decisive.

The part of the load to be applied externally, ext F_{target} can be produced effectively through internal force circulation using steel mobile loading

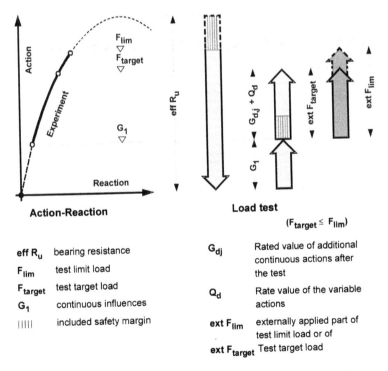

eff R_u	bearing resistance	G_{dj}	Rated value of additional continuous actions after the test
F_{lim}	test limit load		
F_{target}	test target load	Q_d	Rate value of the variable actions
G_1	continuous influences		
‖‖‖	included safety margin	ext F_{lim}	externally applied part of test limit load or of
		ext F_{target}	Test target load

Figure 7.1 Illustration showing the safety concept with load tests on solid structures.

equipment (Figure 7.2). This solution enables high controllable test loads, is versatile in use and fail-safe due to the application of hydraulic presses for creating the test load rather than forces produced by masses with the associated risk of collapse (Steffens, 1995a).

7.3 Requirements, conditions and experience

The most important fundamental requirements for practical applications are:

- structural components which, as application objects, are basically suitable for experiments
- a versatile, controllable loading device
- a computer-aided measurement system with immediate graphical display of the measurement results on the screen for reliable recognition of the test limit load, and
- experienced personnel with a very broad training (civil engineering + measurement techniques + data processing).

Furthermore, the following conceptional prerequisites must be fulfilled:

- Preliminary examinations (with regard to materials and computation)

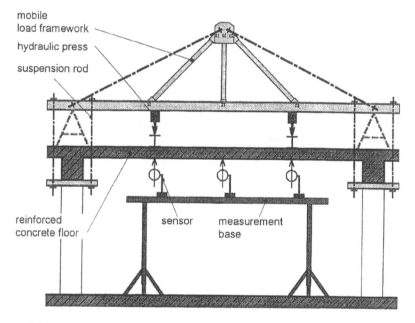

Figure 7.2 Test application (principle).

- Planning of the test programme and the test technique taking into account the safety measures needed
- Definition of the test limit load criteria and their acquisition by measurement
- Keeping of a complete test log.

After the evaluation of the loading tests, the results must be presented in a manner understandable by all parties involved in the project.

- Proof of plausibility by comparison of computational and experimental results
- Explicit details of the permissible reaction effects (live load)
- Description of the requirements for future exploitation of the experimentally determined permissible reaction effects
- Details on the maintenance and monitoring measures needed.

Based on the experience of about 170 projects carried out by the Institute for Experimental Statics at the University of Applied Sciences, Bremen, the following qualitative conclusions can be made. The problem cases of structural analysis considered by the Institute can be differentiated as follows:

20% are hopeless cases needing demolition and new construction or conventional repair methods
40% can be verified using appropriate conventional computational methods after exact preliminary examinations
20% are not economically worthwhile, because the detailed experimental analysis costs do not form a favourable ratio to the expected reduction in total costs
20% show a positive technical and economic result
<1% turn out to be negative due to undetected, severe building faults. This low level of faulty results is due to careful preliminary examination/selection of the projects and must not be attributed to the systematic superiority of load testing compared to static computation.

7.4 Application examples

The following matrix of listed examples highlights the versatility of the practical application of experimental methods in civil engineering. It points to the complexity of the problems which can occur with the retention of existing constructional capital including requirements in the care and protection of monuments (Table 7.1).

Table 7.1 Examples for application of load tests

Reason	Building construction	Civil engineering	Bridge construction
Increase in pay load	32	14	8
Conversion	13	7	1
Building damage	5	6	6
Constructional faults	6	3	1
Guarantee work	8	5	4
Monitoring	15	6	1
Development	6	3	1
Hybrid statics	3	7	7
Monument protection	26	2	3
Total	114	53	32

Source: Institute for Experimental Statics: Experimental assessment of the load-bearing capacity and service ability of structures (number of projects from 1980 to 2000)

7.4.1 Coblenz barracks (Rhine): conversion

In order to be able to use the barracks, built in 1935, for civil purposes, the live load had to be increased from $q = 2.0 \, kN/m^2$ to $q = 3.5 \, kN/m^2$ (auditorium). Computational verification for the reinforced block floor did not prove successful. Load tests on the intervening floors conducted on a random sampling basis proved successful (Steffens, 1995b) (Figures 7.3, 7.4, 7.5 and 7.6).

7.4.2 Secondary school at Wittlich (Mosel): additional floors extension

The former roof floor had to be tested for the higher service load while school operation continued (Figures 7.7 and 7.8).

7.4.3 Valve factory in Bad Homburg: increase in pay load

The storage of material in a valve factory required an increase in the permissible live load for the reinforced concrete joists from $q = 30 \, kN/m^2$ to $q = 50 \, kN/m^2$. Structural component reactions measured during the test: steel strains, concrete compression strain, acoustic emission, crack widths, deflections (Figures 7.9 and 7.10).

7.4.4 Balconies in Coblenz (Rhine): unsafe load-bearing capacity

The four to nine-storey terraced houses are fitted with rows of balconies. Investigatory drillings were taken on the balconies on a random basis. The

Figure 7.3 Former barracks.

Figure 7.4 Mobile loading frame for ext F = 750 kN.

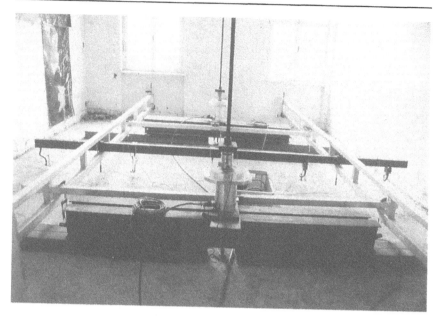

Figure 7.5 Loading gear and measurement bases for displacement measurement.

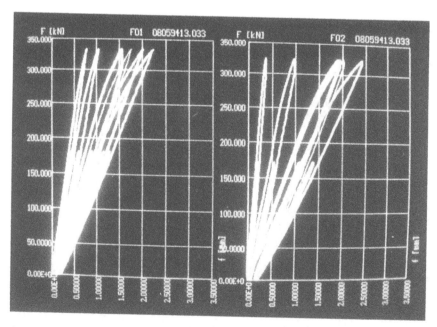

Figure 7.6 Monitor image of the force/deformation curves.

Figure 7.7 View of the building with the added level.

Figure 7.8 Loading gear and measurement equipment on the old roof floor.

Figure 7.9 Creating ext F_{target} = 2600 kN using four loading frames.

Figure 7.10 The joists span l = 11.7 m.

position and condition of the static reinforcement was found to vary significantly and to a large extent it did not produce adequate theoretical load-bearing capacity.

For the alternative experimental assessment of load-bearing capacity, a special, easily fitted and adaptable loading device was developed, which needed no modification of the balcony itself (Figures 7.11, 7.12 and 7.13).

7.4.5 Goods distribution centre: high bay warehouse

The central warehouse was to be equipped with high bay racking with column loads $F \leq 300$ kN. A rough estimate of the computed load-bearing capacity of the hall floor showed permissible column loads of $F \simeq 150$ kN, with the consequence that the foundations of most of the columns would

Figure 7.11 Experimental set-up.

Figure 7.12 Lever arm and measurement base.

have to be strengthened with drilling piles, with negative effects such as a long construction period, disturbance to the operation, dirt and significant costs (Figures 7.14 and 7.15).

7.4.6 Warftenkirche Campen/East Frisia

To maintain the durability of the cross-ribbed vault, tie rods were employed at the height of the imposts. The deformation stiffness of the foundations, pillars, vault and roof timbers could not be estimated and, together with the material creep and the slip in the steel and anchorage construction, a reliable estimate of the prestressing forces could not be obtained by analytical computation.

The prestressing force is controlled via hydraulic presses in relation to

Figure 7.13 Load hook and trestle.

the displacement measurement of the corresponding vault crown (vertical) and the impost (horizontal) in order to moniter the lifting of the vault and hence the restoration of its partially reduced supporting function (Keil and Steffens, 1999) (Figures 7.16 and 7.17).

7.4.7 Crypt at St. Ulrich's Church in Rastede: preservation of monuments

The column shafts of the crypt in gypsum mortar exhibit damage with radial longitudinal cracks. The actual state on the column exhibiting the highest degree of damage was tested by relief measurement (raising the head of the column) using strain gauges. After sawing and regrouting of the head kerf, the load eccentricity was removed (Figures 7.18 and 7.19).

The lifting equipment with four hydraulic jacks surrounds the chapter

Figure 7.14 Stores building before installation of the high bay racking.

Figure 7.15 Producing the test load with a 54 t mobile crane.

Figure 7.16 Cross-ribbed vault with installed test equipment.

Figure 7.17 Control of the prestressing force in the tie rods in relationship to the vault-crown displacement.

Figure 7.18 A gypsum column in the crypt, damaged due to longitudinal cracks.

for the purpose of supporting the column load, followed by regrouting of the head kerf. Electrical strain gauges record the column shaft compression strain during the relief jacking process (Keil and Steffens, 1999) (Figures 7.20 and 7.21).

7.4.8 Reichstag Building, Berlin: pile foundations

The 3,000 pinewood piles in the pile foundation deteriorated in the sap wood after a temporary reduction in the level of the ground water. A series of pile tests was to provide information about the current pile loading capabilities in order to avoid expensive and time-consuming restoration of the foundations (Figures 7.22, 7.23 and 7.24).

On account of the strong flow of ground water, a pit with sheet piling was made in the 2 m thick base using limestone masonry and brick-

Figure 7.19 Lifting equipment for the column under investigation.

chipping concrete so that the original building foundation and the pile heads were free for compression tests.

The test load could be produced hydraulically against loading frames which were in turn reactively anchored via bonded tie rods against the base mass. The loading device was moveable in both directions and designed for test loads of up to 1200 kN. The load/displacement curves for the wooden piles and the plate loading tests were recorded on-line and printed out (Figure 7.25).

7.4.9 Reichstag Building, Berlin: vault

Within the wider scope of building modifications, a computational analysis of the load-bearing capacity of the historic vault on the ground floor was

Figure 7.20 Strain gauges mounted on the column header for measuring the relief deformation.

carried out. It shows impermissibly large eccentricity values for the vault compression as a consequence of half-sided vault loading.

In order to find the actual load-bearing capacity of the construction for new vault loadings and the given load-bearing member including all the real boundary conditions, an experimental determination of load-bearing capacity was carried out. Loading frames anchored against the pillars on the ground floor produced four resulting loads which were converted into four separate controllable partial area loads via tie rods and a load distribution system (Steffens et al., 1997) (Figures 7.26 and 7.27).

7.4.10 Hindenburg lock, Anderten/Hanover

The cross-beams, their bolts and the cast-iron basin niches were to be verified for increased stay forces in all tensile directions in order to withstand the increased loads due to larger barges (Figure 7.28).

The cross-beams were hydraulically tested by an arc-shaped device with a continuously adjustable pulling direction. The maximum test load is F = 350 kN (Figures 7.29, 7.30 and 7.31).

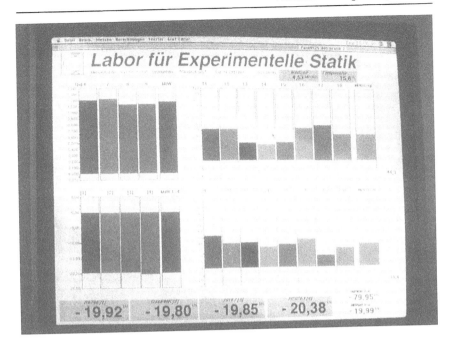

Figure 7.21 On-line display of the measurements.

Figure 7.22 Main facade after removal of the internal structure.

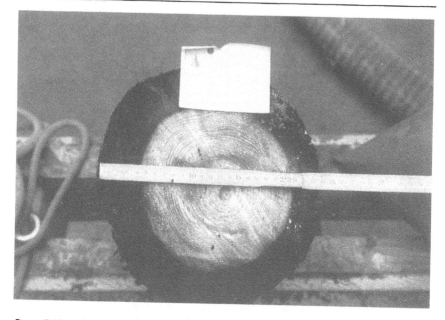

Figure 7.23 Damage due to rotting on the foundation piles.

Figure 7.24 Excavation for the exposure of the test pile heads.

Figure 7.25 Mobile loading framework for F = 1200 kN with side reaction anchorage in the base of the structure.

Figure 7.26 Load application with loading frames anchored against the pillars.

Figure 7.27 Sector-by-sector load distribution equipment.

Figure 7.28 The currently inoperative basin of the double lock.

Figure 7.29 Cross-beam in the basin wall with stay abrasion.

Figure 7.30 Arc-shaped loading equipment and electrical measurement equipment.

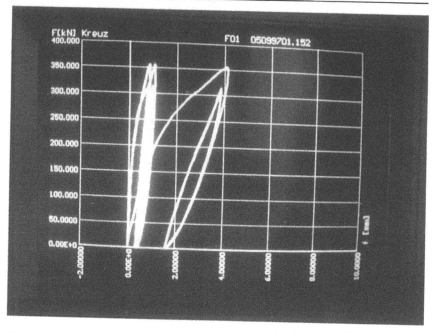

Figure 7.31 Online display of the measurements.

7.4.11 New Museum, Berlin

Due to various reasons (load-bearing system, condition, historical material), the load-bearing capacity of historical floors in the New Museum in Berlin could not be sufficiently classified and exploited using pure computational methods. Supplementary experimental examinations and basic tests were conducted to check the computational examinations, for estimating the condition of the individual load-bearing members and to clarify the load-bearing mechanism in the built-in state (Figure 7.32).

To save weight, the intervening floors had been brick-built using fired clay pots with very thin walls, forming one and two-way curved honeycomb shells (Figure 7.33).

A system identification of the four-section barrel vault was obtained using sampling tests. Then the existing, preventatively higher-dimensioned test set-up was used to carry out an experimental verification of load-bearing capacity for live loads of p = 3.5 kN/m² in order to produce a basis for comparative tests after refurbishing the vault ceiling with topping concrete (Figures 7.34 and 7.35).

Figure 7.32 New Museum, Berlin.

Figure 7.33 Barrel vault of clay pots with temporary roof.

Figure 7.34 Loading device for the sampling tests.

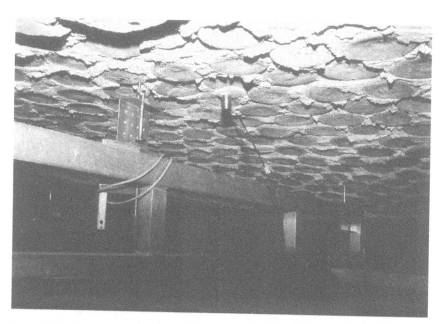

Figure 7.35 View from below the barrel vault with displacement transducers.

7.4.12 ZERI Pavilion, EXPO 2000, Hanover

For EXPO 2000 at Hanover the ZERI organisation erected a pavilion. For the building materials bamboo (guadua augustfolia, chusque), wood (aliso, arboloco) and concrete were used for the load-bearing components and steel bolts and steel fishplates for the joints (Figures 7.36 and 7.37).

The load-bearing capacity and serviceability were verified by computation, where possible based on available material parameters. Due to the large number of unknown factors (material parameters, geometrical deviation, eccentricity, joints, production), supplementary experimental verification of the main components of the load-bearing structure was indicated. These were carried out on a prototype at Manizales (Columbia) and on the finished pavilion in Hanover (Figures 7.38 and 7.39).

The objective of implementing this method of construction with its high demands on the design, construction, ecology and economy using modern technology has certainly been achieved.

7.5 Conclusion

If computational verification does not give realistic results due to inadequate or missing building documentation, complex load-bearing behaviour or obvious or hidden defects, then, after appropriate preliminary investigations, experimental assessments of load-bearing capacity can supply information

Figure 7.36 View of the pavilion.

Figure 7.37 Gallery floor.

Figure 7.38 Cantilever roof test-mass using paving bricks.

Figure 7.39 Gallery test mass using water ballast.

about the real structural behaviour with the inclusion of all existing conditions. In these cases, the experimental investigations usually produce more favourable results than the static computation, as has clearly been shown using some examples. The new method opens up opportunities for retaining building structures, particularly for buildings subject to historical preservation.

References

Anon. Guideline for load tests on solid structures, Part 1 Buildings, Version March 1999 (in German); DafStb, AA test procedure, UA 'Concrete structures' (not published).

Keil, S. and Steffens, K. (1999) 'Experimental assessment of load-bearing capacity on structures subject to monument protection', Conference publication, *Structural Studies, Repairs and Maintenance of Historical Buildings VI'*, Witpress, 23–33.

Steffens, K. (ed.). (1995a) 'Experimental assessment of load-bearing capacity *in situ* for the retention of capital investment or for change of usage', Final Report of the Co-operative Research Project EXTRA II (in German); Bremen. University of Applied Science.

Steffens, K. (1995b) 'Structure retention – reusing barracks (in German)', Bundesbaublatt 44, issue 2, 114–16.

Steffens, K., Wolters, P. and Malgut, W. (1997) 'Experimental assessment of load-bearing capacity on the Reichstag Building in Berlin', Bautechnik 74, issue 7, 434–42.

Index

Milton Keynes UK
Ingram Content Group UK Ltd.
UKHW020031071024
449327UK00032B/3019